The Original Organics Cookbook

ISBN 978-81-7993-155-4

Editor: Nasima Aziz
Managing Editor: Madhu Singh Sirohi
Editorial Coordinator: Roshni Sengupta

Art Director: Priyabrata Roy Chowdhury
Concept and Design: Archana Singh
Photographer: Kuntal Kumar
Image Researcher: Yukti Garg

Published by
TERI Press
The Energy and Resources Institute
Darbari Seth Block
IHC Complex
Lodhi Road
New Delhi 110 003
India

Tel. 2468 2100 or 4150 4900
E-mail teripress@teri.res.in
Fax 2468 2144 or 2468 2145
Web www.teriin.org
India +91 • Delhi (0) 11

Printed on recycled paper

Printed by Multiplexus (India), Delhi

The Original Organics Cookbook

by

Kuntal Kumar

Recipes for healthy living

teri

The Energy and Resources Institute

Contents

About TERI

The Energy and Resources Institute (TERI) is a not-for-profit, research institute with focus on energy, environment and sustainable development.

TERI has undertaken a number of sustainable development initiatives in the hill region of Uttarakhand to improve the quality and quantity of agricultural produce and to improve the socio-economic status of the farmers of that region.

One such initiative has been set up at Supi Farm which is at a height of 7,100 feet in the district of Nainital. A herbal garden has been established to assess the suitability of various culinary, medicinal and aromatic plants under local conditions. The saplings of various species of herbs from the herbal garden are being provided to the farmers so that they will have the confidence to grow these herbs on a commercial basis using bio-fertilisers, and get an additional source of revenue.

The agro-climatic conditions of Uttarakhand are well suited for the cultivation of herbs, and water requirement is very low, leading to luxuriant growth even in dry seasons. The minimal water requirement is met through rain-water harvesting schemes.

TERI provides support to farmers not only in procuring the initial planting material but also by giving information about best farm practices and capacity building. These initiatives are aimed towards helping the local farmers to diversify their crops and to develop village-based micro enterprises.

TERI is supplying culinary herbs to leading hotel chains in India as well as supermarkets and retail stores.

Why Organics?

- Eating organic foods reduces potential health risks. Over 400 different pesticides are used in non-organic farming (conventional farming). Pesticides are poisons designed to kill living organisms and their residues are present in the foods we eat.
- The United States Environmental Protection Agency (EPA) considers 60% of all herbicides, 90% of all fungicides, and 30% of all insecticides as potentially cancer causing. Many EPA-approved pesticides were registered long before extensive research linked these chemicals to cancer and other diseases.

- Switching to an organic diet from a conventional one provides a relatively simple way for parents to reduce their children's exposure to several pesticides as children receive up to four times more exposure than adults to cancer-causing pesticides in foods because of their size and their developing bodies. Baby foods grown with pesticides show pesticide residues.
- Organic farming bans artificial fertilizers which pollute our waters.
- Organic agriculture preserves healthy soils and prevents loss of biodiversity which is critical for a balanced ecosystem for our planet.
- Conventional farmers have ignored the value of preserving a diversity of seed varieties and this has negatively affected soil quality.
- The world can exist quite easily without man but it cannot exist without ants, bees, and other creatures.
- Organic agriculture supports sustainable practices in farming– this is the future need of our planet. Organic agriculture saves energy because it uses more human energy and less fossil fuel.
- Conventional agriculture prefers to selectively breed only those varieties which have fast growth and high production– this does not apply to organically grown food so it is likely to be more nutritious and with more anti-oxidants.
- Research suggests that pesticides and herbicides actually thwart the production of chemicals that act as a plant's natural defence.
- Organic foods are not genetically modified.
- Since World War II, conventional agriculture has destroyed wildlife. Organic farming works with nature to nurture and protect wildlife.
- In conventional farming farm workers have higher exposure to harmful pesticides whereas workers on organic farms are not exposed to this hazard.
- Many people prefer organic food because they say it tastes better. A number of top chefs choose organic foods for the pure taste of organics.
- Organic food sales help small farmers. Although more and more large scale farms are making the conversion to organic practices, most organic farms are small independently owned-and-operated family farms of less than 100 acres and represent one of the few viable strategies for the survival of rural communities.
- Eating organic foods supports a true economy. Conventional food prices do not reflect hidden costs borne by taxpayers, including subsidies, pesticide regulation and testing, hazardous waste disposal and environmental damage.
- Organic foods meet high standards and organic certification is the public's best assurance that their food is grown and handled according to strict sustainable procedures.

Foreword

The Organic Tryst: a Memoir

People express themselves through their inherent talents – some through painting, or music or dramatic skills and some through the food they prepare. Food has been the epicentre of human activity from time immemorial.

This book was conceived sometime in the year 2001 when I started my tryst with organic food courtesy TERI (The Energy and Resources Institute) located at Supi.

The visit to the organic farmland in Supi, a place close to Nainital, was a case of serendipity. Each day was replete with the memories of my childhood when I was closest to nature. The entire cultivation was well laid out in neat rows of terrace farming, meticulously drilled in modern-day organizational concepts. The soil was unspoilt, like the locals, and the aroma was fresh as if all the herbs were dancing in the wind to welcome me – an experience that is one of its kind.

I have left the valleys far behind, but my tryst with organic food continues. After visiting Supi, I began working with organic products in a focused way, and gained an in-depth understanding of their nature and value. Although I was well established in my career I challenged myself to take a new path and create recipes that would extract the best from organic products. My taste buds learnt to appreciate the difference between non-organic and organic food as my new-found adventure continued.

I have grown very fond of the benevolent people at Supi who are eking out their livelihood in the rough mountain terrain. Their contribution to the maintenance of the balance of Mother Nature must be recognized. They are raising the awareness for safe cultivation practices for pure food for the next generation. I am sure one day they will be in the forefront of this movement called Organics which will pave the way for the sustainable development of Mankind and Nature.

Kitchen Talk

Never has food been so refined and people so adventurous and innovative as in the present day and it augurs well for the future also. It is not just the chefs within the industry but full credit must go to the insatiable connoisseurs and food enthusiasts who demand the best.

It is the health aspect that has awakened worldwide interest in organic food and brought back to us the original, long-forgotten pure taste and flavour.

The beautiful part of cooking is that you don't have to get it right the first time, unlike say, a brain surgeon!

As the song says: "We plan, we toil, we suffer in the hope of what? A camel load of idol's eye, the title deeds of Radio City, the empire of Asia, a trip to the moon? No, no, no, no.......simply to wake just in time to smell coffee, bacon and eggs."

In a book on food the writer conveys the excitement and the reader can taste it! The words for colour and perfume bring the food into your mouth. You get mentally prepared to tame the wild fire in your primitive cave to satisfy your senses – yes, an instinct as old as our origins.

Everyone has a different purpose in life: please follow your own dream. But be aware of the role of food in health and happiness. Through the highest consciousness and awareness all your dreams for yourself and for your family will come true. Taking responsibility for your own life is yours. Once you make that special commitment to yourself, that sincere commitment to change, it becomes so easy and everything automatically starts to open up and change does come.

Right from the beginning, 'organics' for me was simply like my mother – in her life and in her kitchen everything has been organic. The most remarkable thing about my mother is that for the last thirty years she has served the family nothing but leftovers. The original meal has never been found!

Kitchen Garden

Mention kitchen garden and you will see a smile on my face. My house is endowed with a big garden and even though we outsource some of the herbs, we do grow a number of them according to their season. A walk around the garden and the senses are replete with the feast of myriad aromas of herbs and the colour and contrast of different vegetables and fruits.

To watch them grow is a marvellous visual experience and you definitely know that what you are bringing to the kitchen is the freshest and best, directly delivered from the garden.

Like my kitchen garden, my family has also grown and my toddler has a diet of healthy organic food well supplemented with productive physical activities. So I can point you in the organics direction with conviction and with many delicious recipes that could grace your table.

Please accept this INVITATION to a long and healthy life, through the laws of nature and the re-discovery of the philosophy of the ancient sages of India, with the commitment to eat only healthy organic food.

Acknowledgements

This book would not be possible without the loving support of so many people who gave me the opportunity to find a new life along the path of Organics. The following is not a hierarchy since each person made a unique contribution.

First and foremost, to my wife and to my newly born son, for whom I wish a TERIfic healthy life and a sense of gratitude to mother earth who grows the elixir of life.

To my parents who had shown me the importance of unadulterated practices including the growing of organic food in my kitchen garden.

I also wish to express my deepest gratitude to the people associated with this wonderful organisation called TERI who gave me the opportunity to share my culinary beliefs with all of you. Mr. Joshi, Director TERI, has shared the same passion.

Amit, my brother and colleague at TERI whose untiring spirit always lifted my morale. His energy level was contagious as he embarked upon so many assignments with consummate ease.

To the very talented people with whom I have interacted at TERI and elsewhere who contributed to the success of this project behind the scenes.

Last but not the least, to the hardest working people on earth, the farmers and the facilitators at the Supi farm, who are the most unspoilt people inhabiting the entire mountain region of Uttarakhand. I salute their organic endeavour which will show the world a way of preserving and protecting our mother earth for future generation.

To my Editor, Nasima Aziz, whose passion for detail has saved me many a blush as a novice writer.

KUNTAL KUMAR

Juices

Raw fruit and vegetables are rich in vitamins, minerals and enzymes. Juices are pure life-force, rapidly assimilated into the blood stream as they require little digestion. Specific fruits and vegetables have therapeutic properties and can be used to treat minor health conditions.

Apple juice: Stimulates the vitality and strength of the body; mitigates anaemia.

Grape juice: Relieves coughs and migraine; loosens phlegm; a good laxative.

Pineapple juice: Strengthens muscle and tissue.

Watermelon juice: Tones the body; prevents hyper acidity; a good summer chiller.

Pomegranate juice: Improves digestion; relieves sore throat.

Sweet lime juice: Cures respiratory disorders, constipation and acidity.

Carrot juice: Stimulates appetite; relieves coughs; a mild laxative.

Beetroot juice: A very good iron tonic; a mild laxative with cooling and cleansing properties.

Cucumber juice: Good for arthritis and eyes, with cooling and cleansing properties.

Tomato juice with celery: Curative properties for mild heart conditions, gastritis and anaemia.

Exciting effects can be achieved by combining different fruit juices. This may require time, patience and a little practice but a sip of these delicious drinks will tell you it was well worth the effort.

Lemon-Lime Ginger Drink

Makes 1 tall glass

1 organic red apple, cored, sliced
½-inch organic ginger
1 cup ripe organic green grapes
½ organic lime
¼ organic lemon
sparkling mineral water

METHOD

Press the apple and ginger together in a juicer, but use less ginger if you think the taste may be too strong. Remove the grapes from the stem and juice along with the lime and lemon. Pour the juice in a large glass and fill to the top with sparkling water. Serve with ice.

Apricot-Pineapple-Strawberry Shake

Makes 1 tall glass

1 fresh organic apricot, diced
¼ cup crushed organic pineapple
½ organic ripe banana
6 ripe organic strawberries
1 tbsp skim milk powder
1½ cups water
1 tsp olive oil (optional)

METHOD

Process all the ingredients in a blender until thoroughly mixed, and serve.

Tropical Fruit Punch

Makes 1 tall glass

1 organic kiwi, peeled
1 organic orange, peeled, sectioned
½ organic mango, peeled, sliced
sparkling mineral water

METHOD
Process all the fruit in a juicer. Pour the juice in a large glass, fill to the top with sparkling mineral water and serve.

Banana-Strawberry Smoothie

Makes 1 tall glass

6 organic strawberries, frozen
1 organic ripe banana, frozen
1 tbsp skim milk powder
1¼ cups water
1 tsp olive oil (optional)

METHOD
Process all the ingredients in a blender until thoroughly mixed, and serve.

Sauces, Spreads, and Pickles

- Coriander Pesto
- Mango Ketchup
- Roasted Tomato and Mint Salsa
- Banana Spread
- Lemon Pickle
- Apple Spread
- Confit of Lemon or Orange
- Apple Pickle

Coriander Pesto

This fragrant, emerald green sauce is inspired by the famous basil pesto of Italy. It can be used for pastas, or as a spread for breads, or as a dressing for salads.

Preparation Time: 10 minutes
Makes 1 jar

2 cups organic coriander leaves
¾ cup olive oil
¼ cup pine nuts, toasted
6 cloves organic garlic
1 tbsp lime juice
1 cup Parmesan cheese, grated
1 tsp salt

METHOD

Place all the ingredients in a blender and blend until smooth. Remove and store in an airtight container, refrigerate until required.

Per serving: 167 Calories (kcal);
27g Total Fat; (87% calories from fat);
6g Protein; 2g Carbohydrate;
8mg Cholesterol;
988mg Sodium

Mango Ketchup

Serve this unique, fruity ketchup with sandwiches, or as a spread on toast, or with the Chilli-Corn Fritters on p. 33.

Preparation Time: 15 minutes
Cooking Time: 1 hour
Makes 1 jar

4 medium organic mangoes, semi-ripe
4 tbsp red wine vinegar
1 tbsp chopped organic ginger
dash of cinnamon powder
1 tsp salt
½ cup brown sugar
½ cup white wine
½ tsp paprika or *deggi mirch* powder
1 clove

METHOD

Peel the mangoes and discard the seeds. Place the mango pulp in a food processor fitted with a stainless steel blade. Add the remaining ingredients and pulse together.

In a heavy-sided saucepan, cook the mixture over low heat for 1 hour or until well-reduced and thickened. Remove from the heat and allow to cool.

Strain through a fine sieve. Refrigerate for 24 hours before using.

Per serving: 132 Calories (kcal);
1g Total Fat; (3% calories from fat);
1g Protein; 32g Carbohydrate;
0mg Cholesterol;
272mg Sodium

Roasted Tomato and Mint Salsa

In Mexico, ripe red tomatoes are roasted on a hot steel *comal* (similar to a cast-iron skillet or Indian *tawa*) until the skins blacken to make the base of many Mexican salsas. The smoky flavour adds a complexity to the taste of ripe tomatoes. Mixing fresh mint as a counterpoint creates the combination of raw and cooked which is one of the classic taste motifs of Mexican cuisine.

Preparation Time: 1 hour
Makes 1 bowl

6 large organic red tomatoes, tops removed
1 clove organic garlic
2 tbsp organic lime juice
3 tbsp olive oil
2 organic chillies, minced, with seeds
1½ tbsp finely chopped organic coriander leaves
3½ tbsp finely chopped organic spearmint
½ tsp organic lime zest
½ tsp organic orange zest
pinch of salt

METHOD

Put a cast-iron skillet or *tawa* over medium-high heat and place the tomatoes on it; cook them until blackened all over. While still warm, pulse tomatoes with the garlic in a food processor until roughly chopped.

Allow to cool to room temperature and add the remaining ingredients. Mix together and set aside for at least 30 minutes before using.

Per serving: 185 Calories (kcal);
14g Total Fat; (65% calories from fat);
3g Protein; 14g Carbohydrate;
0mg Cholesterol;
31mg Sodium

Banana Spread

Cinnamon and lemon give bananas a glamorous new image – it is up to you to use this spread in imaginative and unusual ways!

Preparation Time: 30 minutes
Cooking Time: 30 minutes
Makes 1 jar

10 ripe organic bananas, peeled, crushed

For the syrup
100 ml/½ cup white corn syrup
1 cup water
2 cinnamon sticks
2 lemons, halved
1 tsp lemon zest

METHOD

Place the bananas in a heavy-based pan and simmer gently. Keep stirring after every five minutes, until bananas are completely mashed to a thick consistency.

Place all the ingredients for the syrup in a pan, bring to a boil then simmer for 10 minutes.

Stir the syrup into the banana mixture; simmer for 10 more minutes, stirring continuously and scraping the bottom and sides of the pan to prevent sticking.

Transfer the spread to an airtight container, cool and store in the refrigerator.

Per serving: 135 Calories (kcal);
2g Total Fat; (3% calories from fat);
3g Protein; 32g Carbohydrate;
29mg Cholesterol;
276mg Sodium

Lemon Pickle

Salt, lemon juice and golden sunshine tenderize whole lemons into melting morsels that make you long for days gone by . . . This pickle can be served with any fish dish or with any Indian meal as an accompaniment.

Preparation Time: 30 days
Makes 1 large jar

20 organic lemons
1 cup lemon juice
¾ cup salt

METHOD

Wash the lemons then wipe well to dry.

Peel the skin of each lemon with a stainless steel peeler. (Note: The peel can be used for making confit, see p. 12, after removing the white pith from the underside of the peel.)

Dip each peeled lemon in lemon juice then coat with salt. Place lemons in an airtight jar and top with the remaining lemon juice.

Try to put the jar once or twice a day in sunlight. Once a week stir the lemons to bring the bottom ones to the top. Use after a month.

Chef's Tip

Use this pickle as a home remedy for digestive disorders.

Per serving: 15 Calories (kcal);
trace Total Fat;
trace Protein;
7g Carbohydrate;
0 Cholesterol;
629mg Sodium

Apple Spread

This honey-lemon spread gives you an apple a day all year round. Make this spread one of your healthy breakfast options.

Preparation Time: 10 minutes
Cooking Time: 40-50 minutes
Makes 1 jar

5 ripe organic apples
50g/¼ cup brown sugar
½ tsp organic lemon zest
2 tsp honey
1 cinnamon stick

METHOD

Blanch the apples, remove and discard the skins, the core and the seeds. Chop the apples roughly.

Place the apples in a heavy-based pan with the cinnamon stick and cook on low heat until they are mushy, 30-40 minutes approximately. Cool the mixture, then blend without water to obtain a thick purée.

Pass the purée through a strainer, put it back in the pan and cook till it reduces and becomes very thick.

Add brown sugar and lemon zest; cook for another 5 minutes. Remove the pan from the stove, stir in the honey, cool and store.

Chef's Tip

Substitute brown sugar with molasses or jaggery.

Per serving: 45 Calories (kcal); trace Total Fat; ; trace Protein; 12g Carbohydrate; 0 Cholesterol; 2mg Sodium

Confit of Lemon or Orange

Only the lemon or orange zest (see below) is used for making this confit – the pulp can be used for juice. This marmalade-like garnish adds pazzaz to any dessert!

Preparation Time: 30 minutes
Cooking Time: 30 minutes
Makes 1 small bowl

5 organic lemons or oranges

Syrup
250g/1¼ cups sugar
200 ml/1 cup water
1 tbsp liquid glucose

METHOD

Syrup
Boil the ingredients of the syrup together for 5 minutes then cool.

Confit
Prepare the pith-free zest as described below; cut zest strips lengthwise into thin julienne strips.

Put 1 cup of water in each of three different pans, bring them to a boil and put the zest in one of them. After 20 seconds of boiling, drain the water and refresh the zest with cold water.

Put the zest into the second pan of boiling water and repeat the above mentioned process twice, not forgetting to refresh the zest, after each boil, with cold water. Drain and pat dry.

Measure 250 ml of the syrup into a steel pan, add the zest and heat slowly; cook for 8-10 minutes or until the syrup has reduced slightly and the zest is cooked but still retains its crunch.

Remove from heat and store the lemon strips in a ramekin/small bowl, and allow to cool. The confit (pronounced kohn-FEE) can be stored in the refrigerator for up to 5 days after which it will start to crystallize.

HOW TO ZEST A LEMON OR ORANGE

Zest, sometimes called rind, is the perfumed outermost skin layer of citrus fruit, usually lemons or oranges. The aromatic oils in citrus zest add great flavour to food.

Wash lemons/oranges thoroughly and wipe dry. Carefully remove the outer layer of peel in wide strips using a small knife or vegetable peeler. If zest strips have any pith (the bitter, white underside of the peel) carefully cut or scrape the pith off with a small knife. Cut zest strips lengthwise into thin julienne strips, or as required.

Apple Pickle

The use of apples in pickle is a fusion innovation – it will soon become your favourite!

Preparation Time: 20 minutes
Makes a ½ litre jar

4 green organic apples
¼ cup mustard oil
1 tsp fenugreek seeds
1 tsp yellow mustard seeds
⅓ tsp asafoetida, diluted with 1 tsp water
½ tsp turmeric powder
2 tsp red chilli powder
salt to taste

METHOD

Wash the apples and cut them into even-sized dice, discarding the core and seeds.

Pour oil in a frying pan, add fenugreek and mustard seeds, and allow them to sputter. Add diluted asafoetida, then the diced apples, and sauté for 2 minutes or until the apples are lightly coloured.

Lower the heat, sprinkle turmeric and chilli powders and salt and sauté for 1 more minute.

Cool and store in an airtight container. It will keep in the refrigerator for approximately 20 days.

Per serving: 93 calories (kcal);
8g Total Fat; (71%calories from fat);
trace protein; 7g Carbohydrate;
0mg Cholesterol; 267mg Sodium

CHAPTER 5 APPETISERS

Appetisers

- Baby Spinach Salad, Grilled Onions, Tomato Vinaigrette
- Slow-roasted, risotto-filled Tomatoes, Garden Peas Sauce
- Zucchini with Yogurt Cheese filling, Beets and Beans Slaw, Sesame Vinaigrette
- Potato and Bean-sprouts Galette, Berry Chutney
- Caesar Salad
- Roast Potato Salad
- Chickpea Flour Pancakes, Lettuce with Onion Dressing
- Chilli Corn Fritters
- Red Beet and Gruyere Salad, Orange Vinaigrette
- Stir-fried Bean-sprouts Salad
- Pickled Beet and Apple Salad

Baby Spinach Salad, Grilled Onions, Tomato Vinaigrette

Smoky grilled onions and gently sour tomatoes take spinach into an exotic new world – never again will anyone say that spinach is boring!

Preparation Time: 30 minutes
Cooking Time: 5 minutes
Serves 6

6 cups organic baby spinach

Tomato vinaigrette
1½ tbsp tomato paste
1 tbsp organic tamarind pulp
½ cup chopped organic shallots or baby onions
¼ cup sherry vinegar
1¼ cups olive oil
¼ tsp kosher salt (see p. 109)
dash black pepper

Grilled onions
1 large organic red onion
1½ tbsp balsamic vinegar

METHOD

Wash and dry the spinach

Tomato vinaigrette
Place the tomato paste, tamarind pulp, shallots and vinegar in a blender and mix well. Slowly add the oil a drop at a time at first, then in a slow but steady stream, until it is all incorporated. Add the salt and pepper.

This dressing can be kept in the refrigerator for 1 week.

Grilled onions
Peel and slice the onion thickly enough so that you can handle the slices and they will not fall apart when you cook them. Rub them with a little olive oil and cook the onions over a charcoal fire or under a hot broiler until soft. (Approximately 5 to 6 minutes per side.)

Toss in a bowl with 1-1½ tbsp balsamic vinegar so that the slices become single rings; they can keep in the refrigerator for 2 to 3 days.

Presentation

To assemble the salad, toss the spinach with enough tomato vinaigrette to coat each leaf, being careful not to over-dress the salad.

Divide into 6 portions on individual plates. Arrange the onions over the salad and serve.

Chef's Tip

1. Always serve salads on chilled plates.

2. Tomato paste can be made at home by cooking puréed tomatoes on low heat until they become dark red in colour and reduced to a paste.

3. Tamarind pulp can be obtained by soaking the preserved (dry) tamarind in an equal amount of water. Remove the seeds and blend the pulp; strain and use.

Per serving: 371 Calories (kcal);
49g Total Fat; (93% calories from fat);
1g Protein; 6g Carbohydrate;
0mg Cholesterol; 140mg Sodium

Slow-roasted, risotto-filled Tomatoes, Garden Peas Sauce

This recipe is such a feast for the eyes with vivid contrasting colours – red, white and green like a walk in the garden!

Preparation Time: 30 minutes
Cooking Time: 8 minutes
Serves 4

Risotto
1 cup Arborio rice
¼ cup olive oil
½ medium organic onion, chopped
8 threads of saffron
½ cup white wine (optional)
2½ cups vegetable stock (see p. 42)
a pinch salt
a pinch black pepper
1 tbsp butter
1 tbsp cream
grated Parmesan cheese

Tomatoes
4 large organic tomatoes
1 tsp chopped organic garlic
½ tsp chopped organic thyme
1 tbsp olive oil
1 tsp salt
freshly crushed pepper to taste

Garden peas sauce
½ cup shelled organic green peas
½ cup vegetable stock (see p. 42)

Presentation
1 small organic baby zucchini
4 young organic celery tips with new leaves
a few organic mustard flowers

METHOD

Risotto
Wipe the rice with a cloth to remove any dust but do not wash since it is necessary to preserve the starch.

Heat the oil in a pan, and sweat the onions until transparent. Add the rice and cook for 2 minutes. Add saffron and wine and cook over medium heat till reduced to half, about 5 minutes. Add ½ cup stock, stir till it is absorbed, then add the next ½ cup and stir again till absorbed. Continue in this manner till all the stock is used up, while stirring constantly.

Cook 15 to 18 minutes, *al dente* (see p. 107). Add seasoning and adjust to taste. Remove the pan from the stove and vigorously stir in butter, cream and Parmesan cheese.

Tomatoes
Using the tip of a knife, remove the green head of the tomatoes and cut a shallow cross on the top.

Bring a deep pan half filled with water to a boil, plunge tomatoes in boiling water for 20 seconds, remove and submerge in ice water. Drain the tomatoes and peel off the skin.

Cut the tomatoes across to just expose the seeds, and scoop out the seeds using a teaspoon. Marinate the tomatoes with garlic, thyme, olive oil, salt and pepper for 10 minutes. Roast in a pre-heated oven (150°C/302°F) for 8-10 minutes. Leave it to cool.

Fill the tomatoes with prepared risotto (use leftover risotto in another meal).

Garden peas sauce
Process the green peas with vegetable stock in a blender to obtain a smooth purée, approximately 3-4 minutes. Transfer the purée to a saucepan and put on low heat. Season with salt and keep warm.

Presentation
Slice the zucchini thinly across the length using a mandoline. Blanch in salted boiling water for 20 seconds. Cool and marinate with a little olive oil, salt and pepper.

Place the risotto-filled tomato near the end of an oval plate, spoon the green pea sauce near the tomato and stretch a long line with the tip of a teaspoon.

Roll the zucchini slices and place three of them on the thin line of the pea sauce, equidistant from each other.

Make four individual servings in this manner, garnish with celery and mustard flowers, and serve.

Per serving: 92 Calories (kcal);
5g Total Fat; (93% calories from fat);
2g Protein; 8g Carbohydrate;
4mg Cholesterol; 261mg Sodium

Zucchini with Yogurt Cheese filling, Beet and Beans Slaw, Sesame Vinaigrette

A nutritionally complete salad which provides iron, protein, vitamins, carbohydrates and essential fatty acids – and pure pleasure to all the senses! Learn the procedure for making a fabulous ingredient called hung yogurt or yogurt cheese (see p. 24) – plan ahead as this takes time.

Preparation Time: 30 minutes (plus time for hung yogurt)
Cooking Time: 8 minutes
Serves 4

Sesame vinaigrette
3 tbsp sesame oil
1 tbsp white vinegar
salt and freshly crushed pepper to taste
2 tsp sesame seeds, toasted light brown

Beet and beans slaw
1 organic red beet, peeled
1 handful tender organic beans
1 tbsp olive oil
salt to taste

Zucchini with filling
1 medium organic zucchini
4 tbsp hung yogurt/yogurt cheese or *tomini*

Presentation
1 organic red tomato, blanched, peeled
a few mustard flowers
a few mint sprigs

METHOD

Sesame vinaigrette
Whisk the sesame oil and vinegar until emulsified, season with salt and freshly crushed pepper and stir in the sesame seeds.

Beet and beans slaw
Slice the raw red beet thinly using a mandoline and shred into julienne.

String the beans and cut diagonally in 1-inch pieces, blanch them in boiling salted water for 3 minutes until tender and bright green. Immediately drain them into ice water to preserve the green colour.

Marinate the beans and beet julienne in separate bowls with olive oil and salt.

Zucchini with filling
Slice the zucchini lengthwise using a mandoline; blanch briefly for 20 seconds in salted boiling water, drain and refresh in ice water.

Spoon the yogurt cheese onto one end of the zucchini ribbon and roll it over to encase the filling completely.

Presentation
Quarter the tomato and remove seeds using a sharp knife.
Cut into thick julienne and keep aside.

Put the beans in the bottom of four cold salad plates and top with the beet shreds.

Place filled zucchini on top of the beets.

Drench the vegetables with the prepared vinaigrette dressing and garnish with tomato julienne, mustard flowers, and mint sprigs.

Per serving: 109 Calories (kcal);
7g Total Fat; (40% total calories from fat),
3g Protein; 10g Carbohydrate;
0mg Cholesterol; 345mg Sodium

HUNG YOGURT/YOGURT CHEESE

Also called *Tomini* (Greek). To make 2 cups hung yogurt, place 4 cups whole milk yogurt (*dahi*) in a square of cheesecloth or muslin, gather into a loose bundle and hang it for the whey to drip out completely. Keep a bowl underneath to collect the drips. Drain about 6 to 12 hours to obtain 2 cups of thick and creamy hung yogurt/yogurt cheese.

Potato and Bean-sprouts Galette, Berry Chutney

Here is an interesting and healthy transformation of the potato patties, *aloo tikkis,* of our childhood. The addition of sprouts makes this patty a healthy gourmet starter or snack, which deserves a grown-up name: Galette.

Preparation Time: 45 minutes
Cooking Time: 8-10 minutes
Serves 2

Berry chutney
2 tbsp frozen blueberries, thawed
2 tbsp frozen raspberries, thawed
½ tsp lemon juice
1 tsp icing sugar

Stuffing
1 tbsp oil for pan-frying
½ tbsp minced organic ginger, scrapped
⅔ cup bean-sprouts
⅓ tsp garam masala powder
¼ tsp salt or to taste
⅓ tsp red chilli powder or to taste (optional)
1 tsp cumin seeds, dry-roasted (see p. 62), coarsely ground

Galette
3 large organic potatoes, boiled
1 tsp salt or to taste
⅓ tsp ground black pepper
¼ cup chopped organic coriander leaves, washed
1 medium organic onion, finely chopped
oil for pan-frying

METHOD

Berry chutney
Put the blueberries and raspberries into a pan and bring to a slow boil. As soon as bubbles appear on the surface, remove the pan from the stove and cool. Stir in lemon juice and icing sugar and keep in the refrigerator until required. This chutney can be kept for 3 days in the refrigerator.

Stuffing
Put the oil in a pan, add the ginger and sauté for 20 seconds; add the bean-sprouts and the rest of the stuffing ingredients. Season and transfer to a plate to cool. Divide the stuffing into 10 equal portions and keep aside.

Galette
Peel the potatoes and mash them roughly to make mashed potatoes. Add salt and pepper, coriander leaves and onions, and knead until properly mixed. Divide into 10 equal portions.

Wash and dry your hands and rub them with a little oil. Take each portion of potato mixture and make a ball. Gently flatten each ball into a round patty, ½-inch thick, place a portion of stuffing in the centre and fold the edges together very carefully so that the stuffing is fully enclosed. Make all the patties in this manner.

Heat 1 tsp oil in a non-stick frying pan over low heat. Slip in the patties, not too many at a time, and pan-fry on both sides till crisp and golden brown, adding oil if required, on very low heat.

Serve the patties hot with the berry chutney.

Chef's Tip

Readymade sweet and sour chutney may be used in place of berry chutney.

Per serving: 148 Calories (kcal); 7g Total Fat; (42% calories from fat); 2g Protein; 20g Carbohydrate; 0mg Cholesterol; 6mg Sodium

Caesar Salad

Legend has it that Chef Caesar ran out of traditional dressing one day and hastily concocted this dressing – an unexpected success that has made his name famous. See Chef's Tip for a vegetarian version. To appreciate the complexity of the dressing, the rest of the ingredients are kept simple – just lettuce and homemade croutons.

Preparation Time: 20 minutes
Cooking Time: 10 minutes
Serves 4

Dressing
2 tbsp preserved anchovies
3 tbsp Dijon mustard
2 tbsp chopped organic garlic
4 egg yolks
½ cup red wine vinegar
1 tbsp Worcestershire sauce
1 cup Parmesan cheese
¼ cup olive oil
½ tsp black pepper
salt to taste

Salad
6 bunches organic iceberg lettuce
24 homemade croutons (see p. 32)

METHOD

Dressing
In a large bowl, grind anchovies to a paste-like consistency. Stir in all the remaining ingredients for the dressing and mix well.

Salad
Tear the iceberg lettuce into large pieces, place in a large bowl and dribble with dressing; toss lightly. Scatter croutons into the salad. Serve immediately.

Chef's Tip

For a vegetarian dressing, replace the anchovy and egg yolk with onion and garlic, as follows: Finely chop 1 organic onion and 4 cloves organic garlic, then blend them to a fine paste. Stir in the rest of the ingredients for the dressing. You can keep this in the refrigerator for 4-5 days taking care to use a dry spoon.

Per serving: 242 Calories (kcal);
84g Total Fat; (94% calories from fat);
8g Protein; 4g Carbohydrate;
84mg Cholesterol;
544mg Sodium

Roast Potato Salad

Potatoes are a very good source of carbohydrate – the energy generated within our body. This salad can be eaten either as an appetizer or as a healthy snack. It can even be served with sandwiches as a very good substitute for French fries.

Preparation Time: 30 minutes
Cooking Time: 15-20 minutes
Serves 6

10 medium organic red new potatoes
2 tbsp olive oil
1 tsp black salt or sea salt
½ cup organic red bell pepper, diced
1½ tbsp organic shallots or baby onions, minced
1 tsp fresh organic thyme leaves
1 tsp fresh organic rosemary leaves
1 tsp freshly ground black pepper
1 tsp paprika or *deggi mirch* powder
½ tsp chopped organic garlic
1 tbsp fresh juice of organic lemon
1 tbsp chopped fresh organic parsley

METHOD

Wash and quarter the potatoes.

Preheat oven to 200°C/400°F.

Heat the olive oil in a cast-iron skillet and add the potatoes and salt. Toss to coat the potatoes well and place in the oven until they are slightly soft (approximately 20 minutes).

Add the peppers, shallots, thyme, rosemary, black pepper, paprika and garlic and toss well again to coat the potatoes evenly.

Return the pan to the oven until the potatoes are soft when pierced with a sharp knife (approximately 10 to 15 minutes).

Remove the potatoes from the oven, transfer to a mixing bowl with a slotted spoon and toss well with the lemon juice and parsley. Adjust the salt and black pepper. (For a zestier salad, ½ tsp cayenne or red chilli powder can be added.)

Serve warm or at room temperature the same day. This salad does not save well.

Per serving: 199 Calories (kcal);
10g Total Fat; (40% calories from fat);
4g Protein; 30g Carbohydrate;
0mg Cholesterol; 325mg Sodium

Chickpea Flour Pancakes, Lettuce with Onion Dressing

Lettuce leaves provide a delicate foundation on which carefully selected layers of different colours, tastes and textures are placed to construct a wonderfully delicious, light and fresh presentation – the onion-pancakes make you sigh with satisfaction.

Preparation Time: 45 minutes (plus time for hung yogurt)
Cooking Time: 10-12 minutes
Serves 4

Onion dressing
¼ cup chopped organic onions
1 tbsp capers
1 tbsp red wine vinegar
1½ tbsp grated Parmesan cheese
1 tbsp Worcestershire sauce
3 tbsp olive oil
½ tsp salt

Pancakes
¼ cup gram flour/chickpea flour (see p. 108)
1 egg
1 pinch baking powder
1 pinch salt
1 tsp olive oil
1 tbsp chopped organic onions
1 tbsp chopped organic basil
¼ cup buttermilk

Presentation
2 cups organic iceberg lettuce
1 cup skim-milk yogurt, hung overnight (see p. 24)
1 dry red chilli, seeded, chopped
1 tsp julienne of organic basil
1 tsp chopped organic garlic
1 pinch salt
24 homemade croutons (see p. 32)
2 cloves organic garlic
Parmesan cheese shavings, to garnish

METHOD

Onion dressing
Blend onions, capers, vinegar, Parmesan cheese and Worcestershire sauce in a blender. Run the machine for 10 seconds until all the ingredients are well pulverized.

Gradually pour in the olive oil, while the machine is still running on low speed, to obtain a smooth emulsion. Season with salt, keep in a non-reactive container and refrigerate till required. This dressing will keep in the refrigerator for 3 days.

Pancakes
Mix together gram flour, whole egg, baking powder and salt to make a batter.

Sauté chopped onions in olive oil with basil and fold into the batter.

Add the buttermilk, mix well but lightly, and leave the batter to rest for 15 minutes.

Pour the batter into a warm non-stick pan and cook well on both sides. Cut 12 small roundels of pancake, each 1-inch diameter, using a round cutter.

Presentation
Wash lettuce well and reserve, covered with a damp cheesecloth: This will keep lettuce crisp and crunchy.

Mix together hung yoghurt, chopped red chilli, basil, garlic, salt and keep aside.

Toast croutons in a salamander and rub with a clove of garlic. Discard the garlic and keep the croutons in a warm place

In a non-reactive steel bowl, toss the lettuce with onion dressing and place the lettuce in the centre of four salad plates.

Top with the pancakes and quenelles of hung yogurt (yogurt cheese) and garnish with homemade croutons and Parmesan cheese shavings.

Chef's Tip

As a variation in taste and texture, replace gram flour with cornmeal flour to make the pancakes.

Per serving: 231 Calories (kcal);
16g Total Fat; (57% calories from fat);
12g Protein; 15g Carbohydrate;
12mg Cholesterol; 369mg Sodium

HOMEMADE CROUTONS

Croutons are a salad staple that can be made from virtually any bread. Chances are you have some bread in the kitchen right now that can be used. This recipe makes about four cups of croutons, but is easily adjusted.

10 slices of white or preferably brown bread, cut into ½-inch cubes
2 tbsp olive oil
3 cloves garlic
salt and pepper

METHOD

Heat the oven to 275°F/135°C

Purée the garlic and oil in a blender or food processor, or mince the garlic as finely as possible and stir it into the oil. If you want to use any other seasoning or herbs, add them to the oil now.

Place the bread cubes in a mixing bowl and toss by hand with some of the garlic oil and a few generous pinches of salt and pepper.

Toss the bread cubes well with one hand and continue adding garlic oil so each cube is coated evenly and lightly. Be careful not to over saturate the cubes with oil.

Arrange the bread cubes on a baking sheet in a single layer and place in the oven.

Bake for about 10 minutes, then toss with a spatula and return to the oven. Repeat after another 10 minutes, tasting a cube this time. The croutons should be partly crispy and beginning to turn golden.

Continue to cook and toss periodically until the croutons are golden and crispy throughout. Taste them as they cook to make sure.

Chilli Corn Fritters

Fritters are a deep-fried snack food requiring a lot of fat. My theory is: If I am going to eat fat, I will 'shoot the moon' and choose something as crisp and delicious as fritters! The stiff egg whites and cornstarch make them very crisp and light. I won't even guess the calories! Serve immediately with your favourite sauce, Tomato Salsa or Mango Ketchup (see p. 6)

Preparation Time: 35 minutes
Resting Time: 1 hour
Cooking Time: 10 minutes
Serves 6

Batter
2 eggs, separated
1 bottle carbonated soda
1 cup salty biscuit crumbs
black salt or sea salt, to taste
1 cup cornstarch
1½ cup frozen corn kernels, thawed
1 small organic onion, finely chopped
1 organic green chilli, finely chopped
2 tbsp organic coriander leaves, finely chopped
1 tsp ground cumin
½ tsp paprika or *deggi mirch* powder
1 tbsp lime juice

cooking oil, 3-inch depth in frying pan, or, see your deep-fat fryer directions

METHOD

Batter
Combine egg yolks and 1 cup of the carbonated soda; set aside.

Mix biscuit crumbs, salt and cornstarch and set aside.

Mix corn, onion, chilli, coriander leaves, cumin, paprika and lime juice. Add egg yolk-soda and mix thoroughly. Stir in the biscuit-crumb mixture, adding more soda if the batter is too thick to stir. Cover with plastic wrap, and refrigerate for 1 hour.

Heat the oil and prepare paper towels over parchment (grocery bags work too) on a sheet pan.

While the oil is heating, beat the egg whites to stiff peaks. Fold gently into the fritter batter.

Drop the batter, by the tablespoonful, into the oil and fry until golden brown. Turn with a mesh scoop or slotted spoon. Drain thoroughly on the paper towels. Serve hot.

Chef's Tip

1. Fritters (also called beignet) will hold in a 350°F/180°C oven as long as it takes for the next batch to be cooked. Don't try to make them ahead and hold them too long, because they begin to soften on the inside after about 20 minutes.

2. Variations: Replace the corn kernels with small pieces of cooked fish or shrimp to make Seafood Fritters.

Per serving: 321 Calories (kcal);
6g Total Fat; (17% calories from fat);
7g Protein; 57g Carbohydrate;
75mg Cholesterol;
338mg Sodium

Red Beet and Gruyere Salad, Orange Vinaigrette

The otherwise sedate red beet is like a brother-in-arms with Gruyere cheese or any other semi-hard cheese. When fresh orange juice is added a miracle takes place, like a sunrise!

Preparation Time: 20 minutes
Resting time: 30 minutes
Cooking Time: 60-70 minutes
Serves 4

Salad
3 organic red beets (beetroots)
4 cups assorted organic salad leaves
2 organic pears
¼ cup diced Gruyere cheese
3 tbsp chopped walnuts
few sprigs organic basil leaves, to garnish

Orange vinaigrette
50 ml olive oil
25 ml orange juice
¼ tsp mustard paste
a pinch cardamom powder

METHOD

Orange vinaigrette
Blend the ingredients of the vinaigrette together.

Salad
Boil red beets with skin for 45-60 minutes or until tender. Allow to cool before handling, then peel and cut into finger shapes.

In a non-reactive bowl combine red beets with 3 tbsp of the orange vinaigrette and leave to rest for 30 minutes.

Pick and wash the salad leaves and place equally on four salad plates. Slice the pears and arrange on and around the salad leaves. Scatter the Gruyere cheese and walnuts on top.

Arrange the red beet fingers around the salad leaves and drizzle the salad with the vinaigrette. Garnish with basil leaves and serve immediately.

Per serving: 190 Calories (kcal);
16g Total Fat; (68% calories from fat);
6g Protein; 11g Carbohydrate;
14mg Cholesterol; 63mg Sodium

Stir-fried Bean-sprouts Salad

This low calorie salad could either be a breakfast staple for those people aspiring to remain slim and healthy or it could be a useful option for a healthy snack.

Preparation Time: 25 minutes
Cooking Time: 3-5 minutes
Serves 4

1 tsp sunflower oil
1 tsp chopped organic garlic
1 tsp chopped organic ginger
1 organic carrot, cut in julienne
2 organic capsicum (green bell peppers), cut in julienne
1½ cups moong bean-sprouts
2 tsp soy sauce, low sodium
1 tsp chopped organic coriander leaves
2 organic spring onions or scallions, cut in julienne
1 tsp lime juice
salt and pepper to taste

METHOD

Heat the oil in a wok; add ginger and garlic and cook for 10 seconds.

Add carrots, capsicum, and bean-sprouts and toss on high heat for about a minute. Add soy sauce and coriander leaves.

Add spring onions and lime juice and toss well, check seasoning and serve warm.

Per serving: 99 Calories (kcal);
7g Total Fat; (59% calories from fat);
3g Protein; 8g Carbohydrate;
0mg Cholesterol; 110mg Sodium

WHAT'S A JULIENNE?

Julienne (verb) is a technique used to cut vegetables and other ingredients when you want to heighten their presentation, making them into the size and shape of match sticks or julienne (noun).

- Start by cleaning your cutting board and knife.
- Peel the skin from the vegetable if necessary.
- Trim away any root or stem of the vegetable.
- If the vegetable is round like a potato or carrot, cut it in half and lay it cut side down on the board. This will keep it from rolling.
- Cut the edible part of the vegetable into slices about ¼-inch thick. Cut around the seeds if necessary.
- Cut these slices into even strips ¼-inch thick. The ideal julienne cut is about 2 inches long – any longer will make it hard to get into your mouth!
- Save any scraps for vegetable stock or to use in other recipes.

Pickled Beet and Apple Salad

The sweet and sour tones of this bright-as-the-day salad will tantalize your taste buds!

Preparation Time: 20-30 minutes
Cooking Time: 5 minutes
Serves 4

2 Granny Smith apples
4 small beets
1 tbsp canola oil
1 medium red onion, halved, thinly sliced
3 tbsp lemon juice
2 tbsp drained capers
2 tbsp sour cream
rock salt to taste
freshly ground black pepper to taste
2 tbsp finely chopped flat-leaf parsley, to garnish
1 tbsp chopped fresh chives, to garnish

METHOD

Quarter the apples, remove the cores and slice thinly.

Boil the beets, peel and slice thinly.

Heat the oil in a large skillet over medium heat. Add the onions and apples and sauté for 3-5 minutes or until the apples are soft.

Transfer to a bowl and sprinkle with lemon juice. Add the beets and capers and toss well.

Lightly whisk the sour cream, add to the salad, tossing to coat. Season with salt and pepper.

Transfer the salad to a serving bowl, garnish with parsley and chives and serve.

Per serving: 103 Calories (kcal);
6g Total Fat; (48% calories from fat);
1g Protein; 13g Carbohydrate;
2mg Cholesterol; 87mg Sodium

Soups

- Vegetable Stock
- Carrot Ginger Vichyssoise
- Hot and Sour Mushroom Soup with Bean Curd
- Tomato-Coriander Broth
- Pumpkin Bisque with Lime-Ginger Cream

Vegetable Stock

Any vegetables, vegetable trimmings, herbs and spices are boiled in water till the water becomes flavoured – this is the stock. It can be used in soups, lentils, and vegetable curries to make a flavourful base.

Preparation Time: 15 minutes
Cooking Time: 30 minutes
Makes 12-15 portions

3 medium organic onions
5 stalks organic celery
1 medium organic turnip
300 g organic carrots
3 whole cloves organic garlic
3 bay leaves
1 medium organic fennel head, cut into 1-inch dice
1 star anise
3 litres water

METHOD

Peel and quarter the onions.

Cut the celery into 1-inch pieces.

Peel and cut the turnip and carrots into large dice.

Combine all the ingredients in a stock pot. Bring to a boil then reduce to simmer for 30-35 minutes.

Strain stock through a muslin cloth, discard vegetables.

Cool the stock to room temperature, then refrigerate until required.
It will keep in the refrigerator for 4 days.

Chef's Tip

The trimmings of cauliflower, cabbage, onions, coriander roots etc. generated in the home kitchen can be used for making vegetable stock along with some herbs and spices.

Carrot Ginger Vichyssoise

The original vichyssoise is a soup made only with potatoes and leeks, but I have taken the liberty of adding carrots and a few other innovations. However I have maintained the traditional garnish of chives – enjoy this creamy-rich, cold soup any time!

Preparation Time: 35 minutes
Cooking Time: 15-20 minutes
Serves 8

8 medium organic carrots
2 medium organic potatoes
1 organic leek (white part only)
2 tbsp olive oil
1 clove organic garlic, sliced
1-inch knob organic ginger root, peeled, thinly sliced
2 bay leaves
8 cups vegetable stock (see p. 42)
1 tsp black salt or sea salt
½ tsp black pepper, freshly ground (optional)
2 tbsp cream
organic onion chives or garlic chives, to garnish

METHOD

Peel and chop the carrots and potatoes into large pieces. Wash the leek well and slice.

Warm the olive oil in a soup pot and slowly cook the garlic and ginger until soft.

Add the sliced leeks, raise the heat, and cook until they begin to wilt.

Add the carrots, potatoes, bay leaves, vegetable stock, salt, and pepper, and bring to a boil. Reduce the heat and cook slowly until all the vegetables are soft.

Carefully purée the soup in a blender at low speed until very smooth. Return to the pot, add the cream and adjust the salt and pepper.

Place in a storage container and cool to room temperature before refrigerating.

Adjust the seasoning and consistency if needed before serving.
Serve in chilled soup bowls and garnish with snipped chives.

Per serving: 140 Calories (kcal);
6g Total Fat; (54% calories from fat);
7g Protein; 17g Carbohydrate;
26mg Cholesterol; 1071mg Sodium

Hot and Sour Mushroom Soup with Bean Curd

This oriental soup with bean curd (tofu) is a fully balanced complete meal, brimming with wonderful flavours and aromas. Lemon grass and galangal can become an addiction!

Preparation Time: 45 minutes
Cooking Time: 8-10 minutes
Serves 4

3 cups vegetable stock (see p. 42)
1 cup thickly sliced button mushrooms
1 cup oyster mushrooms
¼ tsp chopped organic green chilli, seeded
1 tsp thinly sliced organic galangal
1 piece organic lemon grass, crushed
5 organic lime leaves
1 cup bean curd/tofu, diced into ½-inch pieces
2 tbsp soy sauce
2 organic tomatoes, seeded, chopped
1 tsp chopped organic spring onions or scallions
1 tbsp chopped organic coriander leaves
½ tsp lime juice

METHOD

Put the vegetable stock in a large pan and bring it to a boil; add both kinds of mushrooms, green chilli, galangal, lemon grass and lime leaves.

Simmer for 5 minutes until all the flavours are well blended in the broth.

Add the bean curd and simmer for another 2 minutes.

Add soy sauce, tomatoes, spring onions and coriander leaves. Adjust seasoning.

Add the lime juice and serve hot.

Per serving: 154 Calories (kcal);
3g Total Fat; (19% calories from fat);
10g Protein; 10g Carbohydrate;
3mg Cholesterol; 236mg Sodium

Tomato-Coriander Broth

Not all tomato soups are thick – this light, delicious broth has lots of coriander leaves and a fresh-herbs taste.

Preparation Time: 30-35 minutes
Cooking Time: 10-12 minutes
Serves 4

4 cups organic tomatoes, quartered
¼ cup organic coriander leaves and stems, well washed
1 small knob organic ginger, peeled, chopped
6 cloves organic garlic, chopped
2 cinnamon sticks
½ tsp cloves
2 green cardamoms
2 bay leaves
10 peppercorns, crushed
1 tsp coriander seeds
1 tsp salt
4 cups water

METHOD

Put the tomatoes in a large pan or stock pot.

Keep some coriander leaves aside for garnish, tear the coriander stems and add these to the pan along with the rest of the ingredients; bring to a boil, reduce the heat and simmer for 15-20 minutes.

Pass the stock through a soup strainer, placing it over a clean pan to catch the strained broth. With a small amount of broth blend the residue in the strainer and strain this also.

Bring the broth to a boil and serve hot, garnished with chopped coriander leaves.

Chef's Tip

Variation: In 1 tbsp vegetable oil fry a few curry leaves and a small pinch of asafoetida diluted in water; add to the broth for a South Indian experience.

Per serving: 92 Calories (kcal);
2g Total Fat; (13% calories from fat);
5g Protein; 19g Carbohydrate;
0mg Cholesterol; 579mg Sodium

Pumpkin Bisque with Lime-Ginger Cream

This is a vegetarian version of the popular seafood bisque. My intention is to reproduce the same strong, creamy and hearty texture.

Preparation Time: 40 minutes
Cooking Time: 8-10 minutes
Serves 8

Bisque
1 small (750g approx.) pumpkin, peeled, diced into 2-inch pieces
2 organic red onions, chopped,
1 leek (white part only), chopped
1 celery stalk, peeled, chopped
1-inch knob organic ginger, peeled, chopped
4 cloves organic garlic, chopped
3 cinnamon sticks
a pinch grated nutmeg
8 cups vegetable stock (see p. 42)
2 tbsp chopped organic parsley, to garnish

Lime-ginger cream
2 tbsp ginger juice (see p. 48)
5 tbsp water
3 tbsp cream, whipped
1 tbsp lime juice
salt and pepper to taste

METHOD

Lime-ginger cream
Put the ginger juice in a pan, along with the water, and simmer for 5 minutes to obtain ginger extract; strain. Add the ginger extract to the whipped cream. Add lime juice, salt and pepper.

Bisque
Roast the pumpkin in a perforated tray in a preheated oven for 20 minutes.

Transfer the pumpkin to a pan and add onions, leek, celery, ginger, garlic, cinnamon and nutmeg. Cover with the stock and cook until tender. Cool for 5 minutes.

Process the bisque in a blender. Strain and return to the pan; heat to warm thoroughly.

Pour the bisque into 8 bowls and decorate each serving with a swirl of lime-ginger cream, with parsley scattered generously on top.

Chef's Tip

Ginger juice. To extract the juice of fresh ginger wash and peel the ginger and grate coarsely. Wrap the grated ginger in a small square of muslin or cheesecloth and squeeze out the juice into a small steel bowl.

Per serving: 154 Calories (kcal);
3g Total Fat; (19% calories from fat);
10g Protein; 10g Carbohydrate;
3mg Cholesterol; 236mg Sodium

Main Course Dishes

- Saffron Risotto with Steamed Vegetables
- Pasta Primavera
- Roast Monkfish with Peas, Baby Chicory, Florentine Onions and Sauterne Velouté
- Olives, Corn and Spinach Cannelloni
- French Beans Sauté
- Grilled Vegetable Lasagne, Mushroom Broth
- Himalayan Bean-sprouts Burger
- Moroccan Vegetable and Chickpea Stew with Couscous
- Bean Curd and Bean-sprouts in light Soy Sauce with Chinese Greens
- Haricot Beans and Bean-sprouts Strudel, Red Pepper Extract
- Cajun-spiced Tofu and Bell Pepper Brochette with Couscous
- Kadhai Paneer
- Prawn Malai Curry
- Butter Chicken (*Murg Makhni*)
- Vegetable Biryani

Saffron Risotto with Steamed Vegetables

This classic risotto is made with starchy, short-grained Arborio rice, wine and vegetable stock. It is stirred constantly, which makes it creamy, whereas the 'basmati rice pulaos' of the East are fluffy rather than creamy.

Preparation Time: 15 minutes
Cooking Time: 25 minutes
Serves 4

Risotto
2 cups Arborio rice
½ cup olive oil
1 medium organic onion, chopped
16 threads of saffron
¾ cup white wine (optional)
5 cups vegetable stock (see p. 42)
a pinch salt
a pinch black pepper
2 tbsp butter
2 tbsp cream
¼ cup grated Parmesan cheese

Vegetables
12 spears organic asparagus
1 medium organic broccoli
1 medium organic cauliflower
1 medium organic zucchini

METHOD

Vegetables
Boil the asparagus spears in salted water for 30-40 seconds. Drain and immediately put into ice water to stop overcooking.

Cut broccoli into florets of natural segments, boil them in salted water for 2-3 minutes until soft. Drain and immediately put into ice water to stop overcooking.

Cut cauliflower into medium florets, cover with water, add a few drops of vinegar to preserve cauliflower's creamy colour and boil for 4-5 minutes or until soft and edible. Drain and refresh in ice water to stop overcooking.

Cut zucchini into fingers, blanch and drain.

Risotto

Wipe the rice with a cloth to remove any dust but do not wash since it is necessary to preserve the starch.

Heat the oil in a pan, and sweat the onions until transparent. Add the rice and cook for 2 minutes. Add saffron and wine and cook over medium heat till reduced to half, about 5 minutes. Add ½ cup stock, stir till it is absorbed, then add the next ½ cup and stir again till absorbed. Continue in this manner till all the stock is used up, while stirring constantly. Cook 15 to 18 minutes, *al dente* (see p. 107). Add seasoning and adjust to taste.

Remove the pan from the stove and vigorously stir in butter, cream and Parmesan cheese. Serve hot with steamed vegetables.

Chef's Tip

Arborio rice is grown in Italy and has short, fat grains and high starch which gives risotto its requisite creamy texture.

Per serving: 784 Calories (kcal);
38g Total Fat; (42% calories from fat);
14g Protein; 102g Carbohydrate;
20mg Cholesterol; 1217mg Sodium

Pasta Primavera

In Italian, *primavera* means 'spring vegetables' which are tossed generously into this pasta. Usually penne is favoured, but any other kind of pasta would give the same delicious results. I have not used the traditional garlic sauce, instead I have introduced garlic into the lightly sautéed vegetables.

Preparation Time: 30 minutes
Cooking Time: 10 minutes
Serves 6

Pasta
3 cups dried pasta (penne or macaroni)
half-filled pot of water, for cooking the pasta
2 tbsp olive oil
1 tsp minced garlic
½ tsp crushed red pepper flakes
2 cups sliced fresh organic mushrooms
3 medium organic tomatoes, coarsely chopped
1 tsp lemon zest
¼ cup chopped organic flat-leaf parsley
salt and freshly crushed pepper to taste
2 tbsp coarsely grated Parmesan cheese, to garnish

Vegetables
2 medium organic carrots
1 medium organic cauliflower
1 medium organic broccoli, cut into florets
1 cup organic asparagus
½ cup organic shelled garden peas

METHOD

Vegetables

Peel the carrots, cut into ½-inch dice; blanch in salted boiling water for 3-4 minutes or until they are cooked *al dente* (just soft enough so that they still must be chewed). Drain.

Cut cauliflower into florets and boil in lightly vinegared water (to preserve its creamy colour) for 4-5 minutes or until soft and edible. Drain and refresh in ice water.

Cut broccoli into florets and boil in salted water for 2-3 minutes until soft. Drain and refresh in ice water.

Cut asparagus into 3-inch pieces; blanch in salted boiling water for 30-40 seconds. Drain and refresh in ice water.

Blanch garden peas in salted boiling water for 1 minute, drain and refresh in ice water.

Pasta
Heat the pasta water to a hard boil, add a pinch of salt and the pasta, and cook until the pasta is *al dente.* Fresh pasta will take 3 to 5 minutes; dry pasta will take up to 15 minutes.

While the pasta is cooking, heat half the oil in a large skillet and cook the garlic and pepper flakes for 1-2 minutes over medium low heat. Turn the heat up to medium high, add the mushrooms and tomatoes, and sauté for 3 more minutes, continually moving them in the pan with a wooden spoon.

Add all the blanched vegetables, lemon zest and parsley, toss to mix, and cook until vegetables are heated.

Drain the cooked pasta in a colander and add to vegetables, toss to mix well, season with salt and pepper.

Serve in individual heated pasta bowls, garnish each serving with Parmesan cheese, and serve immediately.

Chef's Tip

1. To make Farmers Market Pasta, use whatever vegetables are fresh and locally grown.

2. To make Winter Vegetable Pasta, use winter vegetables such as various roots, winter squash, or Brussels sprouts.

3. Use whatever fresh vegetables are available, keeping in mind the colours and textures of the vegetable mixture.

Per serving: 501 Calories (kcal);
24g Total Fat; (43% calories from fat);
18g Protein; 54g Carbohydrate;
13mg Cholesterol; 377mg Sodium

Roast Monkfish with crushed Peas, Baby Chicory, Florentine Onions and Sauterne Velouté

Finding all the ingredients to make this classic dish could be a real challenge – but I have given advice about using locally available substitutes without compromising the results.

Preparation Time: 35-40 minutes
Cooking Time: 45 minutes (excluding tomato petals)
Serves 4

Fish
4 pieces monkfish or any freshwater fish
3 tbsp olive oil
salt and pepper to taste
lemon juice to taste

Florentine onions
8 organic baby onions or shallots
1 orange, zest only
12 coriander seeds
¼ cup grapeseed oil
1¼ cup water
8 peppercorns
1 bay leaf
1 celery stalk
pinch of saffron

Crushed peas
½ cup frozen peas
1 tsp olive oil
1 organic baby onion or shallot, chopped
3g mint, chopped
salt and pepper to taste

Sauterne velouté (sauce)
1 tbsp olive oil
2 tbsp butter
1 organic baby onion or shallot, sliced
10 organic button mushrooms
1½ cups, sauterne or sweet white wine
1 tsp chicken bouillon powder (Knorr)
1½ cups chicken stock
¾ cup double cream

Baby chicory
2 tbsp castor sugar
2 organic chicory, halved
1 tbsp butter
1 cup chicken stock

Presentation
4 organic white asparagus, halved
8 organic tomato petals (see p. 58)

METHOD

Florentine onions
Put all the ingredients, except the onions, in a pan. Bring to a boil then reduce to simmer and place the onions in the liquid. Cook the onions on low heat for 15-20 minutes or until they are cooked through yet firm.

Crushed peas
Bring a small pan of water to a boil, add the frozen peas and cook till done. Drain and refresh in ice water then crush with the back of a steel bowl.
Sweat the onion in olive oil until translucent; add to the crushed peas along with mint. Season with salt and pepper.

Sauterne velouté (sauce)
Heat the olive oil and butter, add the onion and sweat till soft and translucent but without any colour. Add the mushrooms and toss for a couple of minutes or until light golden.

Add the sauterne, chicken bouillon and the chicken stock and simmer for 6 minutes. Add the double cream and simmer for 3 more minutes. Keep aside and rest for 2 minutes. Pass the sauce through a chinois (conical fine mesh strainer), pressing with the back of the ladle to extract as much liquid as possible.

Fish
Heat olive oil in a non-stick pan, place the fish pieces in the pan and cook on one side for 2 minutes or until light golden on the underside; flip over and cook for another two minutes. Transfer the pieces to a roasting tray and roast in the pre-heated oven (180°C/350°F) for 3-4 minutes or until done. Season with salt and lemon juice.

Baby chicory
Heat a cast-iron (heavy) pan, add 2 tbsp castor sugar and continue to heat until a light caramel appears on the surface. Add chicory and toss in the caramel; add 1 tbsp butter followed by the chicken stock and braise for 5-6 minutes or until chicory is thoroughly cooked yet not falling apart; season and keep warm.

Presentation
Blanch the asparagus for 60 seconds, toss in butter, season with salt and pepper and keep warm. Press the green peas in the centre of a plate using a rectangular cutter to outline the shape.

Arrange the tomato petals on the peas, and the asparagus on the sides.

Place the monkfish on top of the petals and stand the Florentine onions in front of the fish.

Froth the sauce with a hand blender and spoon the sauce on and around the fish.

Chef's Tip

Sauterne velouté is a sauce that belongs to the fine dining kitchen. As a substitute, simply sauté 1 chopped onion in 1 tbsp butter until translucent, approx. 2 minutes, and use this for a matching flavour.

Per serving: 390 Calories (kcal);
42g Total Fat; (66% calories from fat);
24g Protein; 18g Carbohydrate;
74mg Cholesterol;
921mg Sodium

HOW TO MAKE TOMATO PETALS

Tomato petals are decorative, delicious and versatile. Use them in salads, sandwiches and main courses.

6 medium ripe organic tomatoes, blanched, peeled
3 cloves organic garlic, sliced thinly
olive oil to brush
icing sugar to dust
a few basil leaves

Preheat the oven to 50°C/122°F.
Quarter the tomatoes, discarding the seeds, and place the tomato segments on a baking tray lined with tin foil.

Brush with oil, place a slice of garlic on each segment, dust with icing sugar and sprinkle torn basil leaves on top.

Put the tray in the oven for 2½ to 3 hours or until each segment has shrunk to almost one-third of its original size, to the size of a petal.

Olives, Corn and Spinach Cannelloni

In this recipe a delectable spinach layer is placed on top of wide strips of pasta (lasagne) which are rolled up like Swiss rolls to make cannelloni tubes. The tubes are then baked in a homemade tomato sauce and served garnished with olives, pine nuts and basil – a mouth-watering combination. The stuffing can vary according to the availability of ingredients.

Preparation Time: 1 hour
Cooking Time: 10 minutes
Serves 4

Pasta
8 lasagne sheets, boiled *al dente*

Stuffing
4 cups organic baby spinach leaves, blanched
2 tbsp virgin olive oil
1-2 cloves organic garlic, finely chopped
½ cup fresh organic corn kernels
salt and freshly ground pepper to taste
2 tbsp shredded organic basil leaves
2 tbsp chopped organic coriander leaves
a pinch grated nutmeg
2 tbsp chopped Kalamata olives
4 tbsp grated Parmesan cheese

Tomato sauce
8 medium ripe organic tomatoes
1 tbsp olive oil
2 cloves organic garlic, finely chopped
salt to taste
1 tsp castor sugar
1 tbsp shredded organic basil leaves

Presentation
1 tbsp pine nuts, toasted
8 Kalamata or any other olives
a few fresh organic basil leaves

METHOD

Stuffing
Squeeze out the excess water from the spinach leaves then chop them finely.

Warm the olive oil in a saucepan over moderate heat, add the garlic, sauté for a minute, then add corn kernels and toss for a minute. Add the spinach and sauté for 30 seconds more; season with salt and pepper.

Remove the pan from the stove, add the basil and coriander leaves, nutmeg, olives and Parmesan cheese. Swirl the mixture with a spoon so that all the ingredients are evenly spread out.

Tomato sauce
Blanch the tomatoes to remove skin; cut them in half and remove the seeds, preserving any juices and flesh in the process.

Warm the olive oil in a saucepan over moderate heat, add the garlic and sauté for a minute. Add the tomatoes with their juice and cook on low heat until almost all the liquid evaporates and they become a thick pulp. Season with salt and castor sugar. Remove the pan from the stove and stir in the basil leaves.

Pasta
Lay a strip of lasagne on a flat surface, place about 1 tbsp of the stuffing on one end, roll over the stuffing like a Swiss roll, securing the other end underneath the roll. Repeat this until you have used all the ingredients.

Using olive oil, grease an ovenproof container large enough to hold all the cannelloni snugly or use 4 individual dishes.

Pour a little of the tomato sauce over the base of the container to cover the base generously. Place the filled cannelloni in the container and cover with more tomato sauce. Cover the container with a sheet of foil that has been greased with olive oil.

Preheat the oven to 200°C/392°F then place the container/containers in the oven and cook till the cannelloni becomes soft. Allow 8 minutes for small containers and 10 minutes for a large dish. Keep the cannelloni warm until ready to serve.

Presentation
Carefully remove the cannelloni from its ovenproof container, and place 2 cannelloni rolls on the centre of 4 plates with the tomato sauce.

Sprinkle with pine nuts and garnish with Kalamata olives and basil leaves.

Chef's Tip

Lasagne sheets, also called Italian pasta sheets, are available with speciality grocers. Please read the cooking instructions on the label as the quality of wheat used may vary from brand to brand, making a difference to the cooking time.

Per serving:
344 Calories (kcal);
38g Total Fat; (66% calories from fat);
24g Protein;
18g Carbohydrate;
74mg Cholesterol;
921mg Sodium

French Beans Sauté

Make a subtle statement with crisp green beans tossed with coriander and fennel seeds. The seeds are dry-roasted to bring out a deliciously toasted flavour (see below).

Preparation Time: 20 minutes
Cooking Time: 10 minutes
Serves 4

450g organic French beans
2 tsp ghee/clarified butter
1 tsp coriander seeds, roasted, powdered
1 tsp fennel seeds, roasted, powdered
⅓ cup chopped organic coriander leaves
½ tsp rock salt
½-inch knob organic ginger, peeled, cut in julienne

METHOD

String and chop the French beans, 3 mm in length. Blanch in salted boiling water for 3 minutes. Drain and put in ice water to stop cooking. Drain again.

Heat the ghee in a pan and sauté the beans for 5 minutes. Stir in the coriander and fennel seed powders and cook for 3 minutes more.

Stir in the coriander leaves and salt.

Check seasoning and serve hot, garnished with juliennes of ginger.

Chef's Tip

1. Blanching the beans helps preserve their colour and their nutritious vitamins.

2. As a variation add grated coconut or well soaked moong lentils or create your own variation.

Per serving: 254 Calories (kcal); 4g Total Fat; (14% calories from fat); 14g Protein; 44g Carbohydrate; 6mg Cholesterol; 270mg Sodium

DRY-ROASTING SPICES, NUTS AND FLOUR

Dry-roasting means to brown ingredients such as whole spices, powdered spices, gram flour and nuts without the use of oil or butter.

Heat a wok or heavy frying-pan till it is medium hot and sprinkle the ingredients on top. Shake the pan or stir the ingredients with a wooden spoon to keep them moving. They are ready when they become highly aromatic and turn slightly darker. Remove immediately to prevent burning.

Grilled Vegetable Lasagne, Mushroom Broth

Slices of assorted grilled vegetables are layered with cheese to create this flourless lasagne-type dish which is both aesthetic and nutritious. It is an excellent vegetarian main course for gluten-allergic people. Portions of vegetable lasagne are placed in a bowl of steaming broth a heady food experience!

Preparation Time: 1 hour
Cooking Time: 35 minutes
Serves 6

Lasagne
4 large round organic eggplants, sliced, grilled
2 green organic zucchinis, sliced, grilled
2 organic baby leeks, white part only, grilled
4 organic red bell peppers, charred, peeled
2 organic yellow bell peppers, charred, peeled
2 medium organic red onions, sliced thinly
5 cloves organic garlic, sliced thinly
3 tbsp grated mozzarella cheese
3 tbsp grated Parmesan cheese
¼ cup fresh organic basil leaves
salt and pepper to taste

Mushroom Broth
1 tbsp olive oil
1 organic baby onion or shallot, peeled, chopped
10 button mushrooms, sliced
5 oyster mushrooms, sliced
2 cups vegetable stock (see p. 42)
salt and freshly crushed pepper to taste
1 tbsp unsalted butter, chilled (used in reheating)

Garnish
8 organic cherry tomatoes, halved
a few sprigs organic parsley

METHOD

Mushroom broth
Place the olive oil in a saucepan, add the onion and sauté for 2 minutes or until translucent. Add the button mushrooms and sauté for another 3 minutes or until they shrink and become lightly coloured. Add oyster mushrooms and toss for 30 seconds.

Pour in the vegetable stock, bring to a boil then lower the heat and simmer for 10 minutes or until the liquid is reduced to two-third. Season the broth with salt and pepper; keep aside.

Lasagne
Preheat oven to 190°C/375° F. To assemble, grease a 12-inch ovenproof dish generously with about 2 tbsp olive oil.

Layer the bottom of the dish with one-third of each of the vegetables: eggplant, zucchini, leeks, peppers, onions and garlic. Sprinkle with about one-third of each of the cheeses and basil and season with salt and pepper. Press down lightly.

Repeat two times to make 3 layers in all, using all the vegetables and cheese.

Cover tightly with foil.
Bake for 25 minutes or until hot and bubbly.

Remove the foil and continue to bake until lightly browned, about 10 minutes. Remove from oven and let set about 10 minutes before cutting and serving.
Reheat the mushroom broth, stir in the chilled butter and whisk vigorously to emulsify. Pour a ladle of broth into a big round bowl along with a good amount of mushrooms.

Cut the lasagne into 3-inch squares. Place in the centre of the bowl of broth.

Garnish
Garnish with cherry tomatoes and parsley sprigs.

Per serving: 196 Calories (kcal);
15g Total Fat; (51% calories from fat);
11g Protein; 21g Carbohydrate;
28mg Cholesterol;
1256mg Sodium

Himalayan Bean-sprouts Burger

This burger is made with finely diced vegetables and bean-sprouts held together with boiled potatoes, which gives it top flavour, bite and crunch. Sandwich it in a toasted bun for peak satisfaction!

Preparation Time: 45 minutes
Cooking Time: 10-12 minutes
Serves 4

1 tbsp olive oil
½ tsp cumin seeds
1 medium organic onion, finely chopped
1 tsp minced organic garlic
¼ medium organic cauliflower, grated, blanched
6 medium organic French beans, blanched, finely sliced
½ cup shelled organic green peas, blanched, crushed
1 organic carrot, blanched, finely diced
4 medium organic potatoes, boiled, grated
¼ cup bean-sprouts
salt and freshly crushed black pepper to taste
oil to pan-fry
4 burger buns

METHOD

Heat the oil in a frying pan, add cumin seeds and when they crackle add onions and garlic and sauté until golden brown.

Add cauliflower, beans, peas and carrots and sauté for 2 minutes.

Add the potatoes and bean-sprouts and cook for another 5 minutes on low heat; adjust the salt and pepper.

Transfer the contents from the pan to a plate and allow the mixture to cool. Form into four patties of equal size and shape. Pan-fry them one by one on a non-stick pan, using a little oil, until golden and crisp from the outside and hot inside.

Cut the burger bun in half, toast under a salamander or on a skillet with some butter, and sandwich the burger in between. Serve with tomato ketchup and English mustard paste.

Per serving: 163 Calories (kcal);
4g Total Fat; (22% calories from fat);
7g Protein; 27g Carbohydrate;
0mg Cholesterol; 153mg Sodium

Moroccan Vegetable and Chickpea Stew with Couscous

Most Moroccan main course dishes are served with couscous, a wheat product, which makes them a complete meal. Although the stew is traditionally made with lamb, I have given you an excellent vegetarian version.

Preparation Time: 1 hour 30 minutes
Cooking Time: 25-30 minutes
Serves 4

Chickpea stew
1 tbsp olive oil
1 small organic onion, chopped
1 medium organic carrot, peeled, diced
1 medium organic zucchini, quartered, cut in triangles
6 organic baby corn, boiled, diced
1 medium organic broccoli, cut into florets
5 cloves organic garlic, minced
½ tsp cinnamon powder
½ tsp turmeric powder
½ tsp cayenne pepper
5 medium organic tomatoes, diced
2 cups vegetable stock (see p. 42)
salt to taste
2 tbsp dried apricots
1 tbsp raisins
1 lemon, zest only
1 cup cooked chickpeas
1 tbsp chopped organic coriander or parsley leaves

Couscous
2 tsp oil
2 small organic onions, chopped
1 small organic carrot, peeled, cut into small dice
1 organic celery stick, peeled, finely chopped
1 cup couscous
2 cups water at room temperature
2 tbsp chopped organic parsley
1 tsp cinnamon powder
2 tsp dry-roasted cumin powder
2 tbsp lemon juice
salt and pepper to taste

METHOD

Couscous
Heat the oil in a saucepan. Stir in onions, carrots and celery. Cook until just tender, about 5 minutes. Stir in couscous, pour in water and cover.

Let rest 5 minutes, till water is absorbed and the couscous becomes soft and edible, then add herbs, spices and lemon juice. Season with salt and pepper. Fluff with a fork.

Chickpea Stew
Heat the oil in a large saucepan over medium heat until hot. Add the onions and carrots and cook, covered, for 5 minutes or until softened.

Add the zucchini, baby corn, broccoli, garlic, cinnamon and turmeric powders, cayenne, tomatoes, stock and salt. Reduce the heat to low, and simmer for 25 minutes.

Meanwhile, soak the apricots in hot water for 20 minutes, then drain and finely chop.

Add the apricots, raisins, lemon zest, and chickpeas to the vegetable mixture and cook 5 minutes longer, or until hot and the flavours are blended. Stir in the cilantro or parsley. Serve this stew with couscous.

Chef's Tip

Couscous is fine granules of semolina wheat. Buy the kind of couscous that does not need steaming but needs only soaking to become edible.

Per serving: 222 Calories (kcal);
4g Total Fat; (13% calories from fat);
5g Protein; 47g Carbohydrate;
0mg Cholesterol; 454mg Sodium

Bean Curd and Bean-sprouts in light Soy Sauce with Chinese Greens

This is the epitome of light and healthy eating in which pure organic vegetables are allowed to sing in their own clear voices, unhindered and uncluttered.

Preparation Time: 30 minutes
Cooking Time: 5-7 minutes
Serves 4

1 tbsp vegetable oil
1 tsp chopped organic ginger
1 tsp minced organic garlic
1 cup diced organic Chinese cabbage
½ cup organic baby corn, halved
1 medium organic yellow bell pepper, deseeded, diced
1 medium organic broccoli, cut into medium florets
1 medium organic zucchini, thickly sliced
¼ cup bean-sprouts
½ cup bean curd/tofu cubes
½ cup thinly sliced Chinese mushrooms
2 tbsp light soy sauce
a handful organic pok choy, washed, dried
1 tsp rice vinegar
1 tsp sugar
1 tbsp sesame oil
2 spring onions, the greens sliced diagonally, to garnish
plain white steamed rice, to serve

METHOD

Heat the oil in a wok. Add ginger and garlic and stir-fry for 1 minute or until fragrant.

Add Chinese cabbage, baby corn, bell pepper, broccoli and zucchini and stir-fry again with some stock or water.

Add bean-sprouts, bean curd/tofu and mushrooms and stir in soy sauce.

Add the pok choy and cook till it wilts slightly in the sauce.

Finish with vinegar, sugar and sesame oil. Check seasoning.

Serve hot, garnished with spring onion greens, with a bowl of steamed rice.

Chef's Tip

The vegetables given in the recipe are just an indication of the range of vegetables that can be used, depending upon seasonal availability, individual preference, cost etc.

Per serving: 117 Calories (kcal); 8g Total Fat; (54% calories from fat); 4g Protein; 11g Carbohydrate; 0mg Cholesterol; 635mg Sodium

Haricot Beans and Bean-sprouts Strudel with Red Pepper Extract

I have experimented with a savoury, healthful version of strudel for you, bursting with texture and taste, and served with fresh, bright vegetables. Amaze your friends and family with this professional-looking, gourmet dish that is not so difficult to make!

Preparation Time: 1 hour
Cooking Time: 15 minutes
Serves 4

Stuffing
1 tbsp ghee/clarified butter or olive oil
2 medium organic onions, sliced thinly
1 tsp sliced organic garlic
1 organic leek (white part only), chopped
2 organic yellow bell peppers, cut in julienne
2 organic red bell peppers, cut in julienne
1½ cups organic French beans, sliced, boiled for 2 minutes
½ cup bean-sprouts
salt and freshly crushed pepper to taste
2 tbsp chopped organic basil leaves

Strudel
4 pieces filo pastry (see p. 106) or spring-roll sheets
1 tbsp ghee/clarified butter
2 tbsp grated Parmesan cheese

Carotene
3 red bell peppers
½ cup water

Vegetables
16 organic asparagus spears
4 organic baby carrots
2 organic zucchinis
4 organic baby corn
1 organic broccoli

METHOD

Stuffing
Heat ghee in a large pan, sauté the onions, garlic and leeks on medium heat for 2 minutes or until fragrant. Add the vegetables in the order given – bell peppers, French beans and bean-sprouts – and toss for another 3 minutes on high heat or until they are partially cooked yet crisp. Season with salt and pepper then transfer the mixture on to a flat tray to cool. When it is completely cool, mix in the basil leaves.

Strudel
Spread out the filo pastry and fold into two equal halves gently. Brush the top layer with ghee. Place the vegetables on the end closer to you, sprinkle with Parmesan cheese, and roll like a Swiss roll. Apply ghee on the other end to completely seal the roll so that it does not open while baking.

Place the strudel rolls on a baking tray, brush them with ghee again and bake in a preheated oven (210°C/410°F) for 10-12 minutes or until the outside is golden brown and crisp; remove from the oven and keep in a warm place.

Carotene
Purée the red bell peppers with ½ cup water, pass through a chinois (conical fine mesh strainer) into a pan. Bring the liquid in the pan to a boil – a red granular froth rises to the top. Reduce the heat to low and skim off the froth – this is the carotene. The rest of the liquid can be used in tomato soup.

Vegetables
Cut the asparagus spears approximately 3 inches from the top. Blanch the asparagus tops in salted boiling water for 30 seconds; drain and refresh in ice water. You can use the bottom part of the asparagus for sauce or soup.

Peel the baby carrots and cut the head to make it flat so that they can stand upright on the plate. Blanch for 2-3 minutes in salted boiling water. Drain and refresh in ice water.

Blanch the baby corn in salted water in which 1 tbsp of vinegar is added to preserve the creamy colour of baby corn. (This applies to all white and cream-coloured vegetables.)

Cut the broccoli into florets and blanch in salted water for 1½ minutes. Drain and refresh in ice water.

Cut one of the zucchinis into half, horizontally, and cut each half into quarters to make fingers. Blanch in salted boiling water for 1 minute, drain and refresh in ice water.

Run the other zucchini through a mandoline to obtain long strands shaped like spaghetti. Blanch the zucchini strands for 30 seconds, drain and season.

Presentation
Place the zucchini strands in the centre of a round plate.

Cut the strudel rolls into three equal pieces and arrange on top of the zucchini.

Reheat all the vegetables in boiling water, drain and season with salt and pepper. Arrange them on all sides of strudel.

Spoon the red pepper carotene around the top and serve immediately.

Per serving: 410 Calories (kcal);
16g Total Fat; (33% calories from fat);
16g Protein; 57g Carbohydrate;
44mg Cholesterol; 275mg Sodium

Cajun-spiced Tofu and Bell Pepper Brochette with Couscous

Here is a dish that is full of freshness and bold international fusion. Brochettes gives you colourful bite-size mouthfuls to be served with couscous which is fine granules of semolina wheat. Buy the kind of couscous that does not need steaming but needs only soaking to become edible. As a variation, try replacing couscous with lemon rice (see p. 77) and enjoy the spectacular results!

Preparation Time: 1 hour (including sauce)
Cooking Time: 12-15 minutes
Serves 4

Brochette
2 organic red bell peppers cut into 1-inch dice
1 medium organic zucchini, halved lengthwise, cut into 1-inch dice
1 block silken tofu, cut into 1-inch dice
12 toothpicks
1 tbsp olive oil
1 tsp Cajun seasoning mix (see p. 77)

Vegetables
6 organic beans, stringed, cut into 2-inch fingers
2 medium organic carrots, peeled, cut into 2-inch fingers

Tomato sauce
4 large organic tomatoes, quartered
1 tsp vegetable oil
1 medium organic onion, chopped
2 cloves organic garlic, chopped
2 bay leaves
2 star anise
salt and pepper to taste
2 tsp shredded organic basil leaves

Couscous
2 tsp oil
2 small organic onions, chopped
1 small organic carrot, peeled, cut into small dice
1 organic celery stick, peeled, finely chopped
1 cup couscous
2 cups water at room temperature
2 tbsp chopped organic parsley
1 tsp cinnamon powder
2 tsp dry-roasted cumin powder
2 tbsp lemon juice
salt and pepper to taste

Garnish
4 green or black olives
a few sprigs organic basil

METHOD

Brochette
Make layers of zucchini, bell pepper and tofu by piercing them in the centre against the tip of a toothpick. Marinate with olive oil and Cajun seasoning mix. Keep aside. Before serving, pan-fry the brochettes in a small amount of oil in a medium-hot non-stick pan until lightly coloured on all sides.

Vegetables
Bring the water to a boil, add salt and blanch the beans for 3-4 minutes or until they are cooked through; drain and immediately put into ice water.

Blanch the carrots in salted boiling water for 4-5 minutes until they are cooked through.
Drain and refresh.

Tomato sauce
Purée the tomatoes in a food processor or blender.

Heat the oil in a medium, heavy-bottomed saucepan and sauté the onions and garlic.
Cook over low heat, stirring often, until the onions are very soft and aromatic, about 6 to 8 minutes.

Add the tomato purée and bring to a simmer over medium-high heat. Reduce the heat to very low, add bay leaves and star anise and let the sauce simmer slowly for 30 minutes, stirring the bottom of the pan often to prevent burning. Season to taste with salt and pepper.

Remove bay leaves and star anise before serving and add the basil leaves. Keep warm.

Couscous
Heat the oil in a saucepan. Stir in onions, carrots and celery. Cook until just tender, about 5 minutes. Stir in couscous, pour in water and cover. Let rest 5 minutes, till water is absorbed and the couscous becomes soft and edible, then add herbs, spices and lemon juice. Season with salt and pepper.

Presentation
Put a portion of the couscous in a medium steel mould and invert in the centre of the plate.

Arrange three brochettes on the sides.

Reheat beans and carrots, season with salt and place on the other side of the plate.

Spoon the sauce around and garnish the couscous with olives and fresh basil.

Chef's Tip

Some people do not like tofu – use *paneer* or other sturdy cheese like mozzarella instead.

Per serving: 289 Calories (kcal);
5g Total Fat; (22% calories from fat);
12g Protein; 35g Carbohydrate;
0mg Cholesterol; 153mg Sodium

LEMON RICE

2 cups boiled rice
6 tbsp oil
½ tsp black mustard seeds
1 tsp split Bengal gram (*chana dal*)
¼ tsp turmeric powder
¼ cup peanuts
3-4 green chillies
a few curry leaves
salt to taste
2 tbsp lemon juice
Heat oil in a pan, add mustard seeds and Bengal gram and allow to splutter. Add turmeric powder and peanuts, fry till golden. Add green chillies, curry leaves and salt and fry for 2 minutes. Remove the pan from the stove, add lemon juice and mix well. Now add this to the boiled rice and mix well.

CAJUN SEASONING MIX

This mix is basically salt with a variety of spices, and is available commercially. If you want to make your own mix, here is the recipe. Experiment with every batch until you have the spice combination you like.
5 tbsp cayenne pepper
3 tbsp black pepper
3 tbsp onion powder
3 tbsp garlic powder
3 tbsp chilli powder
1 tbsp dried thyme
1 tbsp dried sweet basil
1 tbsp dried bay leaf
1 cup salt
Put all the ingredients except salt in a blender and blend to a fine consistency. Don't breathe the dust – it will throw you into a sneezing frenzy! Mix the blended spices with the salt until you achieve a uniform colour. Use as you would salt.

Kadhai Paneer

Kadhai Masala is a well seasoned, thick, tomato-based gravy – one of the basics of Indian cuisine. Once you have the masala ready it lends itself to several variations using vegetarian as well as non-vegetarian ingredients that are speedily tossed together in a *kadhai* or wok. The Basic Kadhai Masala recipe, given below, is followed by a variation using paneer fingers to create an all-time favourite, Kadhai Paneer.

Basic Kadhai Masala
Preparation Time: 25 minutes
Cooking Time: 20 minutes

¼ cup groundnut oil
4 medium organic red onions, peeled, finely chopped
3 tbsp organic ginger-garlic paste
1 tsp turmeric powder
1½ tbsp coriander powder
1 tbsp cumin powder
1 tbsp red chilli powder
8 medium ripe organic tomatoes, chopped
1 tsp dry fenugreek leaves (*kasoori methi*)
salt to taste
1 tsp garam masala powder

METHOD

Heat the oil in a thick-bottomed wok (*kadhai*), or in a deep, round-bottomed pan. Sauté the onions on medium heat, adding 1 tsp salt to release the water, until light golden, approximately 5-6 minutes.

Add ginger-garlic paste and stir for another 2-3 minutes until the raw aroma of the paste disappears and it is well amalgamated with the onions.

With 1 tbsp water make a paste of the turmeric, coriander, cumin and chilli powders, add to the contents of the pan and cook for a couple of minutes.

Add tomatoes and cook until the fat comes to the surface, stirring from time to time to prevent the mixture sticking to the bottom.

Add dry fenugreek leaves and salt then stir in the *garam masala* powder.

Kadhai Paneer

Preparation Time: 20 minutes
Cooking Time: 10-12 minutes
Serves 4

300 g block paneer, preferably homemade
1-inch knob organic ginger, scraped
1 large organic capsicum
3 tbsp ghee
1 tsp cumin seeds
3 dry red chillies
2 cups Basic Kadhai Masala (given above)
½ tsp dry fenugreek leaves (*kasoori methi*)
1 tsp garam masala powder
salt to taste
¼ cup chopped organic coriander leaves

METHOD

Cut the paneer into finger-sized pieces.

Chop two-thirds of the ginger and julienne the rest for garnish.

Cut the capsicum in half, discard the seeds and cut into julienne.

Heat the *ghee* in a wok, add cumin seeds and when they start to splutter, add the dry red chillies and chopped ginger and stir for 30 seconds or until the spices have released their flavour in the oil.

Add capsicum and toss for a further 2 minutes or until partially cooked.

Add the Basic Kadhai Masala and cook until droplets of fat start appearing on the sides, approximately 3-5 minutes.

Add paneer pieces and stir gently for 2-3 minutes until they are well coated; stir in dry fenugreek leaves, *garam masala* and salt and mix well.

Garnish with ginger julienne and coriander leaves; serve hot with your favourite Indian bread – *naan, roti, laccha paratha* etc.

Per serving: 328 Calories (kcal);
11g Total Fat; (29% calories from fat);
27g Protein; 34g Carbohydrate;
29mg Cholesterol;
691mg Sodium

Prawn Malai Curry

Bengal is renowned for its fish and seafood curries and here is the most famous one of all: A delightful combination of prawns with coconut milk, well accented with mustard oil. Ground onions and puréed tomatoes ensure smooth, lump-free gravy.

Preparation Time: 45-50 minutes
Cooking Time: 10-12 minutes
Serves 4

10-12 medium-sized prawns
salt and turmeric powder to rub on prawns
3 tbsp cooking oil, preferably mustard oil
2 bay leaves
2 tbsp organic onion paste
1½ tsp organic ginger paste
1 tsp organic garlic paste
2 tbsp tomato purée
½ tsp turmeric powder
1 tsp chilli powder
½ tsp sugar
salt to taste
½ cup coconut milk
2 tbsp ghee
½ tsp garam masala powder
¼ cup chopped organic coriander leaves, to garnish

METHOD

De-vein the prawns, wash and pat dry. Rub with salt and turmeric powder and keep aside for 10-15 minutes.

Heat the cooking oil in a non-stick frying pan. Fry the prawns till light golden on both sides, remove with a slotted spoon and keep them aside.

Add bay leaves to the hot oil, then add the onion, ginger and garlic pastes and stir for a minute.

Add the tomato purée, turmeric and red chilli powders, sugar and salt and stir till the oil separates and shows at the sides.

Stir in the prawns and mix them well with the masala. Slowly add the coconut milk, then add ½ cup water and bring it to a boil on low heat.

When the curry thickens, approximately 7-8 minutes, add the ghee and garam masala powder, taking care to sprinkle it evenly over the ingredients for best absorption.

Garnish with coriander leaves and serve hot with rice or *naan*.

Chef's Tip

The same curry can be used with other fish and shellfish to make Fish Malai Curry or Lobster Malai Curry.

Per serving: 335 Calories (kcal);
18g Total Fat; (47% calories from fat);
32g Protein; 14g Carbohydrate;
781mg Cholesterol;
1103mg Sodium

Butter Chicken (Murg Makhni)

Luscious Makhni (Butter) Gravy can be considered the king of Indian gravies. Any main ingredient can be combined to give rich results – for example, simmer Paneer in it to make Paneer Makhni.
The Basic Makhni (Butter) Gravy recipe, given below, is followed by its most famous combination with chicken to make Butter Chicken.
This do-it-yourself five-star favourite wins the first prize every time!
It can also be made with cubes of boneless chicken tikkas instead of whole chicken.

Basic Makhni (Butter) Gravy

Preparation Time: 1 hour
Cooking Time: 25-35 minutes

15-20 organic ripe red tomatoes, quartered
2 tbsp chopped organic garlic cloves
10 green cardamoms
1 tbsp red chilli powder
⅔ cup unsalted butter
½ cup single cream
1 tbsp dry fenugreek leaves (*kasoori methi*)

METHOD

Place the tomatoes in a deep pan, add about 4 cups/1 litre water, add the garlic and cardamoms and cook over low heat for 15-20 minutes until most of the tomatoes are reduced to a pulp. Remove the cardamoms and discard.

Transfer the contents of the pan to a blender, allow to cool slightly and blend to obtain a smooth purée.

Strain the purée through a wire strainer, forcing the liquid as much as possible, into a new pan. Discard the residue of skin and seeds.

Slowly bring the purée to a boil then reduce to simmer. Add chilli powder, butter and cream, stir from time to time to prevent sticking to the bottom. Cook until the gravy gets a nice reddish hue, approximately 10-15 minutes. Season with salt and finish with dried fenugreek leaves.

Butter Chicken

Preparation Time: 20 minutes (plus time for hung yogurt)
Marinating Time: 1 hour
Cooking Time: 30-35 minutes
Serves 4

2 medium-sized chickens with skin

Marinade
2 cups whole milk yogurt
1 tbsp organic garlic paste
1 tbsp organic ginger paste
½ tsp garam masala powder
1 tsp paprika or *deggi mirch* powder
½ tsp coriander powder
½ tsp cumin powder
2 tsp lime juice
½ tsp salt

Sauce
3 cups Basic Makhni (Butter) Gravy (given above)
3 heaped tbsp butter
½ tsp paprika or *deggi mirch* powder
½ tsp dry fenugreek leaves (*kasoori methi*), powdered
½ tsp garam masala powder
salt to taste
½ cup cream

METHOD

Cut each chicken into 8 pieces to get 16 pieces.

Marinade
Place the yogurt in a cheesecloth and hang it up to drain out the excess water (see p. 24). Mix drained yogurt with the rest of the marinade ingredients.

Make several gashes up to the bone if using the chicken with bone. Marinate the chicken in the yogurt marinade and leave for at least an hour.

Skewer the chicken pieces and grill in a preheated (180°C/350°F) convection oven for about 15-20 minutes, turning them over midway through the process, until they are nicely brown on the outside and cooked through.

Sauce
In a saucepan, bring the Basic Makhni (Butter) Gravy to a boil, reduce to simmer and add butter; stirring to emulsify in the sauce. Add the chicken pieces and simmer for a further 2 minutes until well coated with the sauce.

Add paprika, dry fenugreek leaves, garam masala powder and salt. Finish with cream just before serving.

Chef's Tip

The chicken can be cooked ahead of time and kept ready for use.

Per serving: 575 Calories (kcal); 4g Total Fat; (42% calories from fat); 72g Protein; 30g Carbohydrate; 212mg Cholesterol; 711mg Sodium

Vegetable Biryani

It is a mistake to think that biryanis can only be made with meat or chicken or fish – try this vegetarian version, to make a handsome, aromatic and classic vegetable biryani using basmati rice and a large variety of vegetables, cooked with onions, spices, tomatoes and yogurt.

Preparation Time: 45-50 minutes
Cooking Time: 30-35 minutes
Serves 4

2 cups basmati rice

Vegetables
¼ cup vegetable oil
1 medium organic cauliflower, cut into medium-sized florets
2 medium organic potatoes, peeled, cut into ½-inch dice
1 medium organic carrot, peeled, halved lengthwise, cut into ½-inch pieces
6 organic French beans, stringed, cut into diagonal diamonds
¾ cup organic shelled green peas
⅓ tsp mustard seeds
2 organic green chillies, sliced diagonally
2 pieces (1-inch each) cinnamon
1 tsp black cumin seeds (*shah jeera*)
4 cloves
½ tsp black pepper powder
3 medium organic onions, finely sliced
salt to taste
1 tsp red chilli powder
4 medium organic red tomatoes, chopped
½ cup yogurt

Garnish
3 tbsp mixed dry fruits (cashew nuts, raisins, almonds etc.)
¼ cup chopped coriander leaves

METHOD

Wash the basmati rice and soak for at least 30 minutes.

Vegetables
Fry the vegetables in medium hot oil, each separately, until lightly coloured.Remove with a slotted spoon and keep aside.

Put 1 tbsp oil in a pan and add mustard seeds, green chillies, cinnamon, black cumin seeds, cloves and black pepper powder and stir for about half a minute. Add the onions and sauté them until they turn light brown, approximately 4-5 minutes. Add salt and red chilli powder and stir.

Add tomatoes and cook until they are reduced to a pulp and well mixed with other ingredients, approximately 5-7 minutes.

Whisk the yogurt and add to the pan, lower the heat and cook for another 30 seconds. Stir in all the fried vegetables and season them.

Rice

Drain the rice and place it in a suitable thick-bottomed pan along with 3¾ cups of water and a little salt. Bring to a boil then simmer until the rice is almost cooked, leaving just a little resistance when pressed, and almost all the water is absorbed.

Add the fried vegetables to the rice and mix very lightly so that the grains remain intact. Cook for another 3 minutes, tightly covered with a lid, on low heat until the rice is fully cooked and elongated, and all the water is absorbed.

Transfer the vegetable biryani to a rice platter. Garnish with dry fruits and coriander leaves. Serve hot with yogurt raita and pickles.

Per serving: 344 Calories (kcal);
7g Total Fat; (18% calories from fat);
62g Protein; 11g Carbohydrate;
331mg Cholesterol;
646mg Sodium

Desserts

- Buttermilk Vanilla Pannacotta with Orange Sabayon and Wild Berry Gratin
- Lemon Grass and Peach Soup, Caramelized Blueberries
- Hot Lemon Grass and Ginger Soufflé
- Holy Basil and Yogurt Cake, Red Currant Compote
- Country Apple and Cinnamon Cake, Vanilla Almond Sauce
- Apple Sorbet
- Honey and Banana Parfait, Homemade Raspberry Sorbet
- Saffron Poached Pears, Berry Sauce
- Mango and Stevia Parfait

Buttermilk Vanilla Pannacotta with Orange Sabayon and Wild Berry Gratin

Italian for 'cooked cream', *pannacotta* is a light, silky dessert served cold, accompanied typically with fruit and sauce. The sabayon is made with egg yolks and cream, like a rich custard sauce. In case you find it difficult to procure the berries, you can use any local seasonal organic fruit available.

Preparation Time: 4 hours
Serves 4

Pannacotta
1¼ tsp gelatine granules
3 tbsp hot water
1 cup cream
½ cup sugar
1 vanilla pod, scraped (see p. 90)
2 cups buttermilk
4 bowls/moulds, 3-inch diameter and 2½-inch height, to set the *pannacotta*

Sabayon
6 egg yolks
¼ cup sugar
¼ cup white wine
1 cup double cream, whipped
2 tbsp Grand Marnier (orange liqueur)

Presentation
2 tbsp organic blackberries, halved
2 tbsp organic blueberries, halved
2 tbsp organic raspberries, halved
4 tbsp organic strawberries, quartered

METHOD

Pannacotta
Melt the gelatine in hot water.

Bring cream, sugar, vanilla seeds and pod to a gentle boil, till sugar dissolves, then remove from heat, remove pod; add gelatine and buttermilk.

Pour into a steel jug, cover the top with cling film to allow the vanilla seeds to spread in the liquid evenly, and let the *pannacotta* rest, approximately 15 minutes.

Pour the *pannacotta* into the bowls/moulds three-fourth to the top and chill in the refrigerator for 2-3 hours or until set.

Sabayon
In the top of a double boiler, combine the egg yolks and sugar, beating well. Add the wine and whisk the mixture constantly until it thickens into fluffy custard, about 5 minutes. Remove from the heat and beat until cool.

Fold the stiffly beaten whipped cream into the cooled sabayon and add the liqueur.

Presentation
Apply heat to the outer surface of the mould, by dipping in hot water to release the panacotta and place it in the centre of a plate.

Arrange the berries on one side, spoon sabayon over the berries, and serve.

In a preofessional kitchen, a blowtorch may be used on the sabayon to give it a light golden/caramel surface before serving – this is what makes it 'gratin' but it is equally good as it is.

Per serving: 352 Calories (kcal);
8g Total Fat; (33% calories from fat);
3g Protein; 35g Carbohydrate;
146mg Cholesterol; 34mg Sodium

HOW TO SCRAPE A VANILLA POD

In some recipes the whole vanilla pod with seeds is used to give an essence that is purer and stronger than the bottled extract. The seeds are left in the food but the pod is generally removed after use. Take a sharp pointed knife, slit the vanilla pod (vanilla bean) lengthways and scrape the seeds out with the tip of the knife.

Lemon Grass and Peach Soup with Caramelized Blueberries

Go all the way in a grand celebration with peach soup mixed with champagne, served in a shot glass, but to be sipped with a spoon. Haunting scents of lemon grass, vanilla, star anise and cloves add to the intoxication. Blueberries swirl temptingly, giving additional sparkle and colour.

Preparation Time: 1 hour 10 minutes
Serves 10

Peach soup
12 ripe peaches
5 cups stock syrup (see p. 109)
2 cups champagne
bouquet garni of a handful basil and mint leaves
2 stems lemon grass, crushed
2 vanilla pods, scraped (see p. 90)
6 star anise
2 cloves
10 peppercorns

Caramelized blueberries
1½ cups blueberries
3 tbsp sugar
3 tbsp water

METHOD

Peach soup
In a pan slowly heat the peaches in the stock syrup and champagne. Add *bouquet garni,* lemon grass, vanilla seeds, star anise, cloves and peppercorns and continue poaching in the liquor for almost an hour or until the peaches become very soft.

Discard the *bouquet garni*, purée peaches using a liquidizer and pass through a *chinois* (a fine conical strainer, or use any fine mesh strainer) into a clean container. Cool the soup and keep in the refrigerator until needed.

Caramelized Blueberries
Make caramel using sugar and water (see below), drop in the blueberries andsimmer gently for 2 minutes; remove the berries from the pan, cool before using.

Presentation
To serve, spoon the blueberries along with their liquid into the bottom of a shot glass, pour in the soup using a sauce boat and stir to allow a part of the blueberry liquid to come to the top.

Per serving: 123 Calories (kcal);
1g Total Fat; (6% calories from fat);
1g Protein; 27g Carbohydrate;
0mg Cholesterol; 4mg Sodium

CARAMEL

In a small, heavy-based saucepan dissolve 3 tbsp sugar with 3 tbsp water.

Heat the pan, bring the liquid to simmer and continue to simmer briskly.

After 3-4 minutes the syrup will start to colour.
Swirl the pan occasionally to ensure an even colour, but don't stir.

Once it is a deep amber colour, remove the pan from the heat and immediately place the base of the pan in a bowl of cold water to stop the caramel colouring more.

Once cooled you will obtain a deep glaze of brown liquid ready for your use

Hot Lemon Grass and Ginger Soufflé

Ginger and lemon grass anchor the ecstasy of molten white chocolate – a dessert to die for!

Preparation Time: 45 minutes
Resting Time: 1 hour
Baking Time: 8 minutes
Serves 5

⅓ cup white chocolate couverture
¼ cup butter
¼ cup icing sugar
4 eggs, lightly beaten
1 tbsp lemon juice
1 tsp lemon grass extract (see p.111)
8 tsp ginger juice (see p. 48)
4 tbsp flour, sieved
multi-grain *tuille* (breadstick)to serve, optional

METHOD

Mix white chocolate with butter in a pan over a double boiler and heat until melted completely.

Stir icing sugar and beaten eggs into the chocolate-butter mixture while still on double boiler.

Remove the pan from the heat and stir lemon juice, lemon grass extract and ginger juice into the batter.

Finally, fold the flour into the batter and rest for at least 1 hour before using.

Butter the inside of a steel bowl that can be used for baking (3-inch diameter, 3-inch height) then lightly dust with refined flour to form a protective coating.

Pour the batter two-third to the top of the bowl and bake in a preheated oven at 190°C/374°F for 8 minutes. Remove the bowl from the oven; rest for 30 seconds; de-mould the soufflé onto a plate and serve hot.

Chef's Tip

White chocolate couverture is a special kind of cooking chocolate with a relatively higher cocoa-butter content which contributes fluidity, smoothness, strength and ease of handling. The high cocoa-solid content heightens the flavour.

Per serving: 373 Calories (kcal);
24g Total Fat; (55% calories from fat);
6g Protein; 36g Carbohydrate;
189mg Cholesterol; 196mg Sodium

Holy Basil and Yogurt Cake, Red Currant Compote

This freezer cake is made with a combination of yogurt, cream, eggs and *agar agar* (vegetarian gelatine, see p. 107). It is unusual for a dessert to have the fragrance of *tulsi*, holy basil (a variety of Indian basil). The fennel in the sugar-free compote continues the experimental theme.

Preparation Time: 45 minutes (plus 3 hours for hung yogurt)
Chilling Time: 2 hours
Serves 4

Yogurt freezer cake
3 cups yogurt
1¾ cups whipping cream
2 egg whites
2 egg yolks
½ cup low fat milk
¼ cup castor sugar
2 tbsp *agar agar*, dissolved in 2 tbsp warm water
5 tbsp light corn syrup
⅓ cup shredded organic basil (*tulsi*) leaves

Red currant compote
½ cup frozen red currants
1 tbsp Equal (sugar-free substitute)
½ tsp fennel powder
1 tbsp powdered dried organic basil leaves, to dust on top

METHOD

Yogurt freezer cake
Place the yogurt in a square of muslin/cheesecloth, gather into a loose bundle and hang it for the whey to drip out completely, for at least 3 hours (see p. 24). You will obtain around 1¾ cups thick and creamy yogurt cheese.

Place it in a non-reactive bowl and whisk until smooth.

Whisk the cream to the soft peak stage in a non-reactive bowl. Keep aside until required.

Whisk the egg whites in a non-reactive bowl, adding 1 tsp of castor sugar to stabilize the foam, until it forms soft peaks. Keep aside.

Put the egg yolks on a double boiler and whisk until light and fluffy. Add the milk and castor sugar and continue whisking until all the sugar is dissolved and the mixture has thickened. Remove from heat.

Add the *agar agar* into the egg mixture and mix well.

Add the corn syrup and stir to incorporate.

Add the whisked yogurt, followed by whipped cream, and fold in lightly until smooth.

Add the basil leaves and mix lightly.

Finally fold in the egg whites lightly until they are well amalgamated.

Pour the mixture into a rectangular mould, 12cm x 10cm, and place in the freezer, minus 12°C, until firm enough to cut and handle.

Red currant compote
Put the red currants in a pan on low heat; add Equal and fennel powder and cook on low heat for a couple of minutes; transfer to a small serving bowl and allow to cool.

Presentation
Cut the cake into squares, 4 cm diameter, dust the top with powdered dried basil leaves, and serve with red currant compote.

Per serving: 480 calories (kcal); 32 gm Total Fat; (58 % calories from fat); 14 g Protein; 37 g Carbohydrate; 223 mg Cholesterol; 195 mg Sodium

Country Apple and Cinnamon Cake with Vanilla Almond Sauce

This moist, mouth-watering cake steams as it bakes in the oven. The combined aroma of cinnamon, vanilla, almonds and apples goes straight to the heart, making this a top favourite.

Preparation time: 1 hour
Baking Time: 60-70 minutes
Serves 8

Cake
¼ cup unsalted butter
1 cup castor sugar
6 eggs
1¼ cup all purpose flour
1½ tsp baking powder
¾ tsp cooking/baking soda
½ tsp cinnamon powder
2 medium apples, cut into 1 cm dice
1 vanilla pod, scraped (see p. 90)

Vanilla almond sauce
2½ cups skimmed milk
½ cup castor sugar
1 vanilla pod, scraped (see p. 90)
2 tsp cornflour
1 tbsp almond flakes

METHOD

Cake
Combine butter and sugar and whisk until creamy, approximately 4-5 minutes.

Add the eggs one at a time and continue whisking until light coloured and well aerated, approximately 6-7 minutes.

Sieve the flour, baking powder, cooking soda and cinnamon powder together, fold into the cake batter with a light hand.

Gradually add the apples and the scraped seeds of vanilla, stir lightly with a spoon to spread the apples and vanilla seeds evenly in the batter.

Pour the batter into an 11x7-inch rectangular baking dish or in a 9-inch square baking dish.

Place the dish in a tray half-filled with water and bake in a preheated (160°C/320°F) oven for about 60-70 minutes or until a toothpick comes out clean when inserted. This type of baking is also known as double-boiler baking where steaming and baking is combined. Remove the dish from the oven and allow the cake to cool down.

Sauce
Bring the milk to a boil with castor sugar and the scraped vanilla pod and seeds.

Add cornflour and stir to allow the mixture to thicken. Add half the almonds, remove from heat, allow it to cool and remove the pod.

Presentation
Slice the cake into even-sized rectangles, 9x5cm. Place on one side of the plate and sprinkle almond slivers on top. Pour the sauce in a shot glass, place next to the cake and serve.

Per serving: 188 Calories (kcal); 11g Total Fat; (54% calories from fat); 8g Protein;
14g Carbohydrate; 163mg Cholesterol; 125mg Sodium

Apple Sorbet

A sorbet contains no animal products so it is totally vegan. Fruit purée and syrup are churned in an ice cream/sorbet machine till frozen and smooth – this one is a vivid green colour, like fresh green apples. In case you do not have a sorbet machine see p.109 for an alternative.

Preparation Time: 20 minutes
Freezing Time: 3 hours
Serves 8

Sorbet syrup
2½ cups water
2 cups caster sugar
⅓ cup liquid glucose

Apples
3 organic green apples, cored
½ lemon, juice only
1 ⅔ cups sorbet syrup
a few drops green food colour

METHOD

Sorbet Syrup
Put all the ingredients for the sorbet syrup in a large pan and bring to boil, stirring frequently to prevent the glucose settling down to the bottom.

Reduce to a simmer for 5 minutes then remove from heat.

Cool it down before using. It can be kept in the refrigerator for up to 15 days.

Apples
Cut the apples into rough dice without removing the skin.

Combine apples, lemon juice and the measured amount of the sorbet syrup in a blender and blend until smooth and almost like paste or pâté.

Pass this mixture through a fine mesh strainer, pressing it with the back of a ladle to extract as much liquid as possible.

Churn it immediately in the ice cream/sorbet machine for 4½-5 minutes or until frozen and smooth.

The sorbet needs to be kept very cold to preserve the vivid green colour.

Per serving: 100 Calories (kcal);
trace Total Fat; (2% calories from fat);
trace Protein; 23g Carbohydrate;
0mg Cholesterol; 12mg Sodium

Honey and Banana Parfait with Homemade Raspberry Sorbet

While this parfait uses fruit pulp, cream, egg yolks and agar agar (vegetarian gelatine, see p. 107), the sorbet is nothing but frozen, churned fruit pulp – a winning combination in this beautifully light and cool dessert.

Preparation time: 45 minutes
Chilling time: 2 hours
Serves 4

Parfait
2 tbsp castor sugar
⅓ cup water
4 egg yolks
3 bananas, puréed
¼ cup organic honey
2 tsp *agar agar*, dissolved in 2 tbsp warm water
1 cup unsweetened whipping cream

Sorbet syrup
2½ cups water
2 cups caster sugar
⅓ cup liquid glucose

Raspberry sorbet
1 kg raspberry purée
350 ml sorbet syrup

METHOD

Parfait
Put the sugar and water in a pan and boil until the sugar has completely dissolved and forms a thick syrup; keep hot.

Whisk egg yolks in a non-reactive bowl until creamy and almost ribbon-like when lifted and dropped.

Add the hot syrup and continue whisking until the mixture has thickened and cooled slightly.

Add banana purée and honey followed by *agar agar*. Stir to mix well.

In a non-reactive bowl, whisk cream until it reaches the soft-peak stage, a stage where it just retains its peak when the whisk is lifted. Fold into the yolk mixture, pour into a rectangular mould, 12x10x4cm, and place in the refrigerator to set.

Sorbet syrup
Put all the ingredients for the sorbet syrup in a large pan and bring to a boil, stirring frequently to prevent the glucose settling down to the bottom. Reduce to a simmer for 5 minutes and then remove from heat. Cool it down before using. It can be kept in the refrigerator for up to 15 days.

Raspberry sorbet
Combine the raspberry purée and sorbet syrup in a bowl and whisk to mix completely.

Churn in the ice-cream/sorbet machine for 5 minutes until frozen and smooth.

Transfer it to an airtight container and keep in the freezer until needed.

Presentation
Remove the parfait from the refrigerator and cut into desired shapes. Serve with the sorbet.

Chef's Tip

1. Other fruit sorbets like mango, pear, strawberry, lychees etc. can be prepared in the same manner using the fruit purée of your choice.
2. For a smooth texture without seeds, pass the berry mixture through a fine mesh strainer.
3. In case you do not have a sorbet machine see p. 109 for an alternative.

Per serving: 262 Calories (kcal);
19g Total Fat; (61% calories from fat);
25g Protein; 2g Carbohydrate;
69mg Cholesterol; 22mg Sodium

Saffron Poached Pears with Berry Sauce

In this miraculous blend of delicate flavours the pears and berries have a sublime relationship, pledged with saffron. The berries are left whole in the sauce and retain their shape and colour because they, like the pears, are only lightly poached.

Preparation Time: 20 minutes
Cooking Time: 25-30 minutes
Serves 5

Pears
4 firm pears
some lemon juice
½ tsp saffron threads
1 tbsp milk, warmed
½ cups white wine or verjuice (optional)
2 cups water
1 cup sugar
1 cinnamon stick

Berry sauce
½ cup assorted berries
1 lemon, zest only
2 tsp icing sugar

Presentation
1 tsp flax seeds, toasted
1 tsp sunflower seeds
fresh mint sprigs

METHOD

Peel whole pears and rub them all over with lemon juice.

Dissolve the saffron in the milk and keep aside.

Place wine, water, sugar, cinnamon stick and saffron milk in a large saucepan. Bring to a boil, stirring occasionally, until sugar dissolves.

Add pears and cover with a piece of butter paper pressed down onto the liquid so that the pears stay submerged. Poach gently for 20-30 minutes or until just tender. Remove pears with a slotted spoon.

Berry Sauce
Take about half a cup of poaching liquid; reduce on high heat until 2 tbsp of liquid is left. Add the berries, lemon zest and icing sugar to the reduced liquid, (do not cook further) and keep aside.

Presentation
Cut the pears into quarters.

For each serving, divide the berry sauce between the 5 plates; arrange the poached pears on top and sprinkle flax seeds and sunflower seeds around. Garnish with fresh mint sprigs and serve.

Chef's Tip

Verjuice is an acidic, sour liquid made from unripe fruit, primarily grapes.

Per serving: 129 Calories (kcal); Trace fat Total Fat; (0.5 % calories from fat); 1g Protein; 26 g Carbohydrate; 0mg Cholesterol; 7 mg Sodium

Mango and Stevia Parfait

Stevia is a herb from which a natural sweetener is obtained (see p. 113) making this dessert a favourite among diabetics. The parfait is served in filo cups giving a classy and elegant look to what is essentially a simple and easy-to-make dessert.

Preparation Time: 45 minutes
Freezing Time: 1 hour
Baking Time: 5-6 minutes
Serves 4

Stevia syrup
20 stevia leaves
1 cup of water

Parfait
4 gelatine leaves or 12g granules or 12g agar agar
4 organic ripe mangoes, pulped
⅔ cup stevia syrup
1 cup double cream

Filo cups
1 piece filo pastry (see p. 106)
clarified butter to brush

Garnish
a few chopped assorted nuts like pistachio, pecan etc.
a few mango slices
a mint sprigs

METHOD

Stevia syrup
Bring the stevia leaves and water to a boil, reduce the heat and simmer for 3-4 minutes. Strain and keep aside.

Parfait
Soak gelatine leaves in cold water for 3 minutes, or, dissolve the gelatine granules or agar agar in 3 tbsp hot water.

Purée the mango pulp until smooth, transfer to a pan and cook on low heat for about 10 minutes or until it thickens; add the stevia syrup and the gelatine (squeeze gelatine leaves to remove extra water) and whisk to obtain a smooth mixture.

Remove from heat and leave at room temperature for 10 minutes.

Lightly whip the cream, fold it into the prepared mixture and freeze the mixture for an hour or so to set.

Filo cups
Cut the filo pastry into two equal halves, brush one with clarified butter and put the other on top. Now cut into four equal pieces and bake in moderately hot oven (160°C/310°F) for 5-6 minutes or until golden. Press the baked filo, one by one, inside a small steel bowl to obtain cups. Allow them to cool before removing to retain the cup shape.

Presentation
Fill scoops of parfait into the filo cups. Sprinkle chopped nuts on the top and sides. Arrange slices of mango on top of parfait and garnish with sprigs of mint.

Per serving: 262 Calories (kcal);
19g Total Fat; (61% calories from fat);
2g Protein; 25g Carbohydrate;
69mg Cholesterol; 22mg Sodium

HOW TO HANDLE FILO PASTRY DOUGH

Filo, also spelled phyllo, is a paper-thin, versatile pastry dough that is used in Greek cooking. From appetizers to desserts, there are thousands of uses for this remarkable product.

To defrost, leave the unopened package in the refrigerator overnight and the filo will be ready to use the next day. Do not thaw at room temperature because the sheets then have a tendency to stick to each other.

Before unrolling the filo, make sure that you have all your ingredients and utensils ready at hand. If you are going to bake the filo pastry immediately, your oven should be preheated.

Once you have unrolled the filo, work with one sheet or a portion of a sheet at a time. Keep the remainder covered with plastic wrap or a damp cloth to prevent the sheets from drying out.

Use a pastry brush to spread an ample amount of butter on the sheet, lightly dabbing it all over.

Glossary of Cooking Terms, Ingredients and Equipments

Agar or *agar-agar* is a white, semi-transparent vegetarian gelatine substitute sold in packages as washed and dried strips, or in powdered form. In Indian cuisine, *agar-agar* is known as china grass and is used for making desserts.

Al dente refers to the desired texture of cooked pasta or cooked vegetables in Italian cuisine. It literally means 'to the tooth' or just soft enough so that it still must be chewed. When the pasta or vegetable is cooked al dente, there should be a slight resistance in the centre when you bite into it. The same term can be used to describe the initial cooking of risotto.

Anchovies are tiny fish generally filleted, salt-cured, canned in oil, and sold flat or rolled. Because they are so salty, anchovies are used sparingly to flavour or garnish sauces and other dishes.

Bean curd see Tofu

Bouquet garni is a bundle of herbs (bay leaf, thyme, parsley, celery, leek etc.) tied together with string and used to prepare soup, stock, and various stews. The bouquet is boiled with the other ingredients and removed prior to consumption.

Brochette refers to food which has been cooked on a skewer, typically in a French style.

Cajun seasoning mix see p. 77

Chinois, a cone-shaped sieve or strainer, is a handy kitchen utensil to have for giving sauces and soups a smooth finish.

Couscous see p. 68

Croutons, homemade see p. 32

Double boiler is a specialized piece of kitchen equipment consisting of two fitted saucepans. The larger saucepan is partially filled with water brought to a simmer or boil. The smaller, inner saucepan uses this indirect heat to melt chocolate, cook custards and sauces, or even melt wax for candle-making. A double boiler can also be improvised with a large saucepan and a bowl for home cooking.

Dry-roasted spices, nuts and flour see p. 62

Filo pastry see p. 106

Gelatine is an odourless, tasteless and colourless thickening agent, which when dissolved in hot water and then cooled, forms a jelly. Gelatine is pure protein derived from beef and veal bones. Granulated gelatine is the most common form of unsweetened commercial gelatine, but it is also available in leaf or sheet form. Sheet gelatine produces a clearer gelled product than its granular counterpart.

Gram flour, a flour made from ground chickpeas, is also known as *besan, kadale hittu*, chickpea flour and garbanzo flour. Gram flour contains a high proportion of carbohydrates but no gluten. It is used to make the batter of the best loved snack-food in Indian cuisine *pakoras*, deep-fried fritters and has many versatile uses in a vegetarian diet.

Grapeseed oil is extracted from grape seeds and has a relatively high smoke point, approximately 420°F (216°C), so it can be safely used to cook at high temperatures for stir-fries, sautéing and fondue. Other positive attributes are its clean, light taste, making it a good choice for salad dressings, or as a base for infusing or flavoring with garlic, rosemary or other herbs or spices.

Julienne, 'matchstick cut' vegetables (see p. 37)

Mandoline is a kitchen utensil consisting of a flat frame with adjustable blades, for slicing vegetables into thin to thick slices. Firm vegetables and fruits (such as potatoes and apples) can be cut with uniformity and precision. On most machines, the food is held in a metal carriage on guides so that fingers are not in danger.

Non-reactive bowls are made of glass, stainless steel, or plastic. They are used for mixing and tossing citrus salads, chutneys etc. A 'reactive' bowl, like aluminium, reacts to acid-based dressings e.g vinaigrette, lemon juice, yogurt, tomato etc, and copper which hastens the browning of fruits and vegetables like banana, apple, potato etc.

Sabayon is a light custard made by beating egg yolks with a liquid over simmering water until thickened and increased in volume. Sabayon may be sweet or savoury, served simply as a sauce, and often flavoured with alcohol.

Salamander is a piece of kitchen equipment used to heat food from above.

Salt is available in different forms. **Table salt** is a fine-grained refined salt with additives that make it free-flowing. **Kosher salt** is an additive-free, coarse-grained salt. **Sea salt** is the result of the evaporation of sea water and comes in fine-grained or larger crystals. **Rock salt** has a greyish cast because it retains more minerals and harmless impurities and comes in chunky crystals. All salt crystals have to be crushed before use. **Black salt** comes in reddish-black crystalline chunks or in powdered form. It has a distinct sharp taste and smell. Both black salt and sea salt are organic.

Sauternes is a type of dessert wine from the Sauternais region of the Bordeaux wine-making area of France.

Sorbet is a frozen fruit dessert made with a sweetener and a fruit purée sometimes flavoured with wine or liqueur. Since it contains no animal products it is classified as vegan. In case you do not have a sorbet machine, whisk the ingredients for sorbet in a suitable container. Put in the freezer, whisk three times after 30 minute intervals. Rest in the freezer until completely frozen.

Stock syrup Put 7½ cups water, 6 cup castor sugar and 1 cup liquid glucose in a large pan and bring to a boil, stirring frequently to prevent the glucose settling down to the bottom. Reduce to a simmer for 5 minutes then remove from heat. Cool it down before using. It can be kept in the refrigerator for up to 15 days.

Tofu/bean curd also known as **doufu,** is made by coagulating soy milk, and then pressing the resulting curds into blocks. It has very little flavor or smell on its own, so it can be used either in savory or sweet dishes, and is often seasoned or marinated to suit the dish. The production of tofu from soy milk is similar to the production of cheese from milk (fermentation).

Tomato petals, baked, for garnish see p. 58

Quenelle in its classical definition is a dumpling poached in water but it can also refer to the typical oval shape of sorbet when scooped using a spoon/ice cream scoop.

Velouté is one of the five classic mother sauces of French cuisine. It is often served on poultry or seafood dishes, and is used as the base for other sauces.

Yogurt, hung yogurt, yogurt cheese see p. 24

Zest, lemon, orange see p. 12

Glossary of Organic Herbs

Basil, sweet basil is a herb of the mint family, extensively used in Mediterranean cooking. Choose evenly coloured leaves with no sign of wilting. Refrigerate basil, wrapped in barely damp paper towels and then in a plastic bag, for up to 4 days. Alternatively, finely chop the cleaned basil, combine it with a small amount of olive oil and freeze in tiny portions to flavour sauces, salad dressings, etc. The Indian variety, called 'holy basil', *tulsi, tulasi*, is a principal herb of Ayurveda, the ancient traditional holistic health system of India.

Chives are the smallest species of the onion family. They are referred to only in the plural, because they grow in clumps rather than as individual plants. There are two types which find their way into the kitchen: onion chives and garlic chives (Chinese chives). The leaves (straws) are shredded for use as garnish for fish, potatoes and soups, and their flowers may also be used as a garnish. It has insect-repelling properties which can be used to control garden pests.

Coriander leaves, cilantro, Chinese parsley, (coriandrum sativum), is an annual herb with bright green, lacy leaves that are broad-lobed at the base of the plant, and slender and feathery higher on the flowering stems. The leaves and young upper stems are used for garnishing, and the thick stalks and roots are used for added flavour in cooking.

Galangal, blue ginger, (Indonesian name *laos*) is a rhizome with culinary and medicinal uses, best known in Thai and other Southeast Asian cuisines. Though it resembles (and is related to) ginger in appearance, it does not taste exactly like ginger. In its raw form, galangal has a soapy, earthy aroma and a pine-like flavor with a faint hint of citrus.

Garlic is a cousin to leeks, chives, onions and shallots. The edible bulb or 'head' grows beneath the ground. This bulb is made up of sections called cloves, each encased in its own parchment-like membrane which is usually peeled before use in recipes. Green garlic is young garlic before it begins to form cloves. It resembles a baby leek with a long green top and white bulb, sometimes tinged with pink. The flavour of a baby plant is much softer than that of mature garlic.

Ginger is a root with a knobbly appearance and a fresh, spicy, pungent fragrance. It is the mainstay of Asian and Indian cooking it can be ground, grated or slivered for use. It is available in two forms young and mature. Young tender ginger has a mild flavour and a pale, thin skin that requires no peeling, whereas mature ginger has a tough skin that must be carefully peeled away to preserve the delicate, most desirable flesh just under the surface.

Leek

Kaffir lime, also known as kieffer lime, *makrut,* or *magrood,* is a type of lime native to Indonesia, commonly used in Southeast Asian cuisine. It is a small, rough, bumpy green fruit that grows on a very thorny bush with aromatic leaves. The leaves can be used fresh or dried, and can be stored frozen. It is well suited to container growing.

Lemon grass has a citrus flavour and is widely used as a herb in many Asian Cuisines. It can be dried and powdered, or used fresh. The stalk itself is too hard to be eaten except for the soft inner part. However, it can be finely sliced and added to food. It may also be bruised and added whole as this releases the aromatic oils from the juice sacs in the stalk. It is commonly used in teas, soups, and curries and for flavouring poultry, fish, and seafood. Lemon grass infused in water can be simmered and reduced to obtain an extract for heightened flavour.

Leek belongs to the same family as onion and garlic and looks like a giant green onion. The edible portion of the leek is the white onion base and the light green stalk. Onion-like layers form around a core which, when tender, may be eaten, but as the leek ages the core becomes woody and generally unusable. The smaller the leek, the more tender it will be. Refrigerate leeks in a plastic bag up to 5 days. Before using, trim rootlets and leaf ends. Slit the leeks from top to bottom and wash thoroughly to remove all the dirt trapped between the leaf layers. Leeks can be cooked whole as a vegetable or chopped and used in salads, soups and a multitude of other dishes.

Mint has bright green leaves, purple-tinged stems and a distinct fragrance and flavour. Choose leaves that are evenly coloured with no sign of wilting. Store a bunch of mint, stems down, in a glass of water with a plastic bag over the leaves. Refrigerate in this manner for up to a week, changing the water every 2 days. Mint is used in both sweet and savoury dishes and in drinks.

Flat leaf Parsley

Onions come in two main classifications, green onions which are immature, and dry onions, which are simply mature onions with a juicy flesh covered with dry, papery skin. Green onions (spring onions) are also called scallions. They have a white base that has not fully developed into a bulb, and green leaves that are long and straight. Both parts are edible. Baby onions are small-sized dry onions and have a milder flavour than full-grown onions. 'Sweat the onions' means to cook chopped or diced onions slowly in a small amount of butter or oil until they release some of their moisture and become slightly translucent.

Parsley is a fresh-flavoured herb used for flavouring and garnishing. The most popular kind is curly-leaf parsley but flat-leaf Italian parsley has a stronger flavour. Choose bright-green leaves that show no sign of wilting. Wash fresh parsley, shaking off excess moisture, and wrap first in paper towels, then in a plastic bag. Refrigerate for up to a week. Parsley is an excellent source of vitamins A and C.

Rosemary is a member of the mint family and its silver-green, needle-shaped leaves are highly aromatic, with a hint of both lemon and pine. Rosemary can be used as a seasoning in a variety of dishes including fruit salads, soups, vegetables, meat (particularly lamb), fish and egg dishes, stuffings and dressings.

Shallots are formed more like garlic than onions, with a head composed of multiple cloves, each covered with a thin, papery skin. The skin colour can vary from pale brown to pale grey to rose, and the off-white flesh is usually barely tinged with green or purple. Shallots are favoured for their mild onion flavour and can be used in the same manner as onions.

Stevia is a shrub in the sunflower family. As a sugar substitute, stevia's taste has a slower onset and longer duration than that of sugar, with its extracts having up to 300 times the sweetness of sugar. Thanks to TERI who pioneered the cultivation of stevia in India, this natural sweetener is a boon to diabetics and others on carbohydrate-controlled diets.

Thyme is a bush with grey-green leaves giving off a pungent minty, light-lemon aroma. It is widely used in cooking to add flavour to vegetables, meat, poultry and fish dishes, soups and cream sauces. It's a basic herb of French cuisine and an integral part of *bouquet garni.*

French Thyme

Notes